Steeling the Mind

Combat Stress Reactions and Their Implications for Urban Warfare

---

Todd C. Helmus, Russell W. Glenn

Prepared for the United States Army

Approved for public release, distribution unlimited

The research described in this report was sponsored by the United States Army under Contract No. DASW01-01-C-0003.

Library of Congress Cataloging-in-Publication Data

Helmus, Todd.
Steeling the mind : combat stress reactions and their implications for urban warfare / Todd Helmus, Russell Glenn.
p. cm.
"MG-191."
Includes bibliographical references.
ISBN 0-8330-3702-1 (pbk.)
1. War neuroses—United States. 2. Urban warfare. I. Glenn, Russell W. II. Title.

RC550.H47 2005
616.85'212—dc22

2004026856

The RAND Corporation is a nonprofit research organization providing objective analysis and effective solutions that address the challenges facing the public and private sectors around the world. RAND's publications do not necessarily reflect the opinions of its research clients and sponsors.

RAND® is a registered trademark.

Published 2005 by the RAND Corporation
1776 Main Street, P.O. Box 2138, Santa Monica, CA 90407-2138
1200 South Hayes Street, Arlington, VA 22202-5050
201 North Craig Street, Suite 202, Pittsburgh, PA 15213-1516
RAND URL: http://www.rand.org/
To order RAND documents or to obtain additional information, contact
Distribution Services: Telephone: (310) 451-7002;
Fax: (310) 451-6915; Email: order@rand.org

Preface

Combat stress casualties can severely limit the manpower required to conduct military operations. This manpower loss may be even more accentuated during urban combat operations. Commanders and NCOs in the U.S. military should develop the necessary skills to treat and prevent stress casualties and understand their implications for urban operations. To impart this knowledge, this monograph reviews the known precipitants of combat stress reaction (CSR), its battlefield treatment, and the preventive steps commanders can take to limit its extent and severity. In addition, this monograph reviews the stress casualty evidence of prior urban battles in order to enhance understanding of the risks of urban operations with respect to the development of CSR. Both treatment and prevention are also examined from an urban operations perspective.

This study will be of interest to military commanders and senior NCOs. The actions of these individuals have the greatest impact on the occurrence of battle fatigue and its proper treatment. This monograph was written with this audience in mind. Military and civilian medical practitioners and scientists will also take interest, especially with regard to the review of stress casualty rates in prior urban operations.

This research was undertaken for the Training and Doctrine Command of the U.S. Army and was conducted in RAND Arroyo Center's Force Development and Technology Program. RAND Arroyo Center, part of the RAND Corporation, is a federally funded research and development center sponsored by the United States Army.

For more information on RAND Arroyo Center, contact the Director of Operations (telephone 310-393-0411, extension 6419; FAX 310-451-6952; email Marcy_Agmon@rand.org), or visit Arroyo's web site at http://www.rand.org/ard/.

Contents

Figures

Tables

Summary

The stress of military operations can tax a soldier to his outermost limits. Negative reactions to this stress may include misconduct behaviors, post-traumatic stress disorder (PTSD), and combat stress reactions (CSR). CSR, the focus of this report, is defined as any response to combat stress that renders a soldier combat ineffective. Symptoms vary but often include debilitating forms of anxiety and depression and the "thousand-yard stare."[1] Physical symptoms are also common. Depleting combat units of critical manpower, wartime rates have ranged from 10 to 30 percent of those wounded in action (WIA), with certain units experiencing rates well above 50 percent of those wounded. Still, rates of CSR can be limited by a variety of prudent leader actions, and the disorder is amenable to front-line treatment.

Urban operations are often characterized by a three-dimensional environment with innumerable fields of fire, poor concealment for offensive forces, close-quarters fighting, diluted leadership, restrictive rules of engagement, and ambiguity as to the identity of hostiles. Despite these challenges, the U.S. military is confronted with an increasing and unavoidable demand to place troops in potentially hostile cities. The unique demands required of this operating environment may pose a significant threat to the psychological makeup of friendly forces. Thus, one goal of this report is to evaluate the

[1] The "thousand yard stare" is generally considered a symptom of a mild form of CSR and is not necessarily indicative of combat ineffectiveness.

risk of CSR in urban operations. In addition, this report details the restorative methods applicable to CSR and steps that may be taken by military commanders and senior NCOs to prevent its occurrence. Treatment and prevention are also viewed from the urban operations perspective.

The major chapters of this report include a look back at the history of forward psychiatry,[2] a review of the factors that precipitate stress reactions, an examination of the risks of CSR in urban combat operations, a study of CSR treatment, and a review of recommendations for its prevention. The summary of each of these chapters is provided below.

History

To fully understand the present-day approach to CSR, it is necessary to review both its history and the individual and environmental conditions that mediate its appearance. The "soldier's heart" and "nostalgia" of the Civil War graduated to the "shell shock" of World War I. The British Army struggled to stem the tide of soldiers succumbing in vast rates to a seemingly physical malady, microhemorrhaging of the brain caused by the explosion of artillery shells and the shock waves they produced. Shell shock, eventually called "war neurosis," was characterized by physical ailments, paralysis, and a host of psychiatric symptoms. The first vestiges of forward psychiatry appeared, in which treatment consisted of an expectation to return to duty, rest, and military drill provided close to the front and relatively soon after onset. The memory aid "Proximity, Immediacy, Expectancy, and Simplicity" (acronym PIES) was subsequently coined to summarize these principles.

Forgetful of French, British, and American hard-earned lessons, the U.S. military in World War II attempted to prevent stress casual-

---

[2] Forward psychiatry is here defined as actions taken to return soldiers suffering from stress-related reactions to their fighting units and to limit the tide of psychologically precipitated evacuations.

ties with a screening program that denied vast numbers of aspiring servicemen the opportunity to serve their country. Meanwhile, soldiers and marines at the battles of the Kasserine Pass and Guadalcanal suffered notoriously high rates of "psychoneurosis." Forward psychiatric treatment principles embodied in PIES were then quickly adopted. The term "exhaustion" was applied to stress casualties in order to depathologize the disorder and to communicate that simple rest was sufficient for restoration. Rates of stress casualties throughout the war varied from 15 to 30 percent of those wounded.

In Vietnam, rates of stress casualties were low, possibly due to relatively brief engagements, rare subjection to indirect fire, and a 12-month rotation system. The rotation policy, however, may have created more problems than it solved in that it hampered unit cohesion and morale and limited combat effectiveness. Shortly after this war, Israel was suddenly attacked by its neighbors in what became known as the Yom Kippur War. Never incorporating the psychiatric lessons learned by the Americans and British, the Israelis suffered high rates of stress casualties, many of whom were lost to long-term disability. By the time Israel launched its invasion of Lebanon in 1982, its armed forces had successfully incorporated forward psychiatric doctrine. While dramatically improving stress casualty treatment, the Israelis still suffered CSR rates close to 23 for every 100 WIA.

The U.S. Army recognized combat stress control as an autonomous Medical Department functional area and distinct Battlefield Operating System. In the Gulf War, mobile psychiatric teams helped limit the rate of psychiatric evacuations during the buildup to the ground war. Subsequently, combat stress control teams have deployed to a number of different locations and helped provide outpatient treatment, command consultations, unit surveys, and stress management classes.

Precipitating Factors

Throughout these various conflicts, the military has learned a number of lessons about the individual and environmental conditions that precipitate CSR. From an individual perspective, while personality has never been directly implicated in CSR, it has been identified as a factor in combat effectiveness and in stress symptoms induced by stressful training regimens. The presence of home-front stressors and low educational levels do show a relatively strong relationship to breakdown[3] as seen in CSR. Variables related to the unit environment are also important. Low levels of unit morale, unit cohesion, and faith in self and command have shown disproportionately higher stress casualty rates than individuals or units without such problems. While elite special operations forces are well protected from CSR, individuals who comprise combat service support (CSS) units, as well as reserve units that come under fire, can experience rates of CSR as a function of total physical casualties well above their infantry counterparts. Battlefield factors that induce stress reactions include combat intensity, initial and prolonged exposure to combat, static warfare, and deficits in sleep, hydration, and nutrition.

Risk of Stress Casualties in Urban Warfare

With this background in mind, we evaluated the degree of stress and rates of CSR for urban conflicts of the past. Soldiers and marines interviewed for this report testify that urban combat is inordinately stressful. Furthermore, it is the view of many medical and scholarly authorities that the stressors of urban combat are likely to increase the risk of CSR. Historical data, however, from the Battles of Brest, Manila, and Hue show no evidence of increased rates of stress casualties. It is suggested that the failure to find high urban-generated CSR rates, despite subjective reports of increased stress, may be due to the sense of control enjoyed by urban fighters who, due to close-quarters fighting, are more able to engage enemy combatants and benefit from the therapeutic effect of the distractions inherent in high-intensity combat. Add to this the fact that all operations reviewed by the

---

[3] In this report, the term "breakdown" is used synonymously with CSR.

authors were offensively fought. It must also be considered that the reviewed battles do not constitute the full range of modern history's urban operations. The battles of Brest, Manila, and Hue were not hampered by the presence of civilian populations who masked non-uniformed combatants and attempted to impede or directly engage U.S. forces. In contrast, many recent-day urban conflicts, especially military operations other than war, are characterized by operating environments replete with civilians, both friend and foe, who pose significant challenges to U.S. forces. Interviews suggest that the civilian element of urban operations may be a key risk factor for the development of acute or chronic stress reactions. Consequently, future research efforts must focus on evaluating the specific and acute psychiatric consequences of this operating environment.

Restoration Methods

The U.S. military must remain vigilant about the psychiatric risks posed by future urban operations as well as operations on other types of terrain. To this end, commanders must be educated about CSR's treatment and prevention. When applied to the actual battlefield setting, CSR has been called battle fatigue. The diagnosis of battle fatigue is complicated by a number of factors, including symptoms that evolve over time and are often characterized by a vast array of behaviors or conditions that combine in disparate ways. To simplify symptom identification, soldiers are cautioned to beware of "persistent, progressive behavior that deviates from a [service member's] baseline behavior." Treatment for battle fatigue can be applied by both formal mental health assets such as combat stress control (CSC) units and division mental health teams and the combat or support unit of the battle-fatigued individual. Treatment by mental health assets first involves a screening to rule out the existence of medical and psychiatric conditions that require alternative treatments. The treatment principles for battle fatigue have frequently been summarized by the four "R"s: *Reassurance* of a quick recovery from a confident and authoritative source; *Respite* from intense stressors; *Replenishment* in the form of water, a hot meal, sleep, regulation of body temperature, and hygiene; and *Restoration* of perspective and confidence through conversation and working.[4] PIES, i.e., proximity (treat close to the front), immediacy (treat as soon as possible), expectancy (assert expectation of recovery), and simplicity (simple treatment approach) are also used to describe battle fatigue treatment. Similar principles are applied when the unit is the treating agent, though the unit's tactical condition may impose significant restraints on caring for the soldier.[5] Importantly, simple and brief restoration techniques should reverse the psychological decline of many battle-fatigued soldiers whose symptoms are mild and identified early.

While PIES or the four Rs as administered by mental health units at locations separate from the soldiers' units successfully return many battle-fatigued soldiers to duty, there are limitations. Rates of return to duty (RTD) vary widely from 15 to 75 percent, and soldiers who have succumbed to acute stress reactions may be at increased risk for subsequent battle-induced stress reactions. Some afflicted soldiers may need to be reassigned to duties that limit involvement in direct combat operations. Soldiers with a prior history of battle fatigue may also be at increased risk for the development of PTSD.

Finally, PIES implemented by mental health units in locations separate from soldiers' units may be ill suited to maneuver warfare in which tactical units move great distances in short periods.[6] Such conditions make it difficult to return soldiers to duty. In addition, some U.S. Army mental health assets currently deploying to Iraq are doing so as organic to maneuver brigades. Changes such as this should continue. Reliance on NCO peer mentors who can coordinate in unit

---

[4] Paraphrased from COL James Stokes, M.D., written comments to the author, November 18, 2004.

[5] Historically, the unit has been the preferred primary treatment agent. External mental health assets such as a psychiatrist and other mental health personnel assigned to the division, or a psychiatric detachment or combat stress control units, serve as a second echelon depending upon the case and the combat environment.

[6] However, preventive mental health teams were able to successfully follow maneuver units during Operation Iraqi Freedom's major combat phase and provide in-unit PIES interventions.

care and act as liaisons with conventional mental health units may also serve an important need.

Prevention

Finally, we describe a number of prudent and common-sense actions available to commanders to limit the occurrence of stress casualties. Although programs geared to screen out individuals who are at risk of psychiatric reactions lack feasibility, soldiers new to their unit can be placed in an indoctrination program that teaches unit history and lore. Training programs should seek to increase confidence and acquaint soldiers with various stressors that lie ahead. The authors review a training program that accomplishes just such a goal. In addition, commanders must make special efforts to enhance unit cohesion, and CSS units should develop combat-specific training programs.

For military operations, the authors suggest a number of specific steps. Combat stress control units can perform a variety of services for commanders and their units, ranging from evaluations of unit morale to seminars on stress and CSR-specific issues. Members of these units, however, must be intimately acquainted with the units that they service. Given the influence that rotation and replacement policies have on cohesion and length of exposure to combat, the authors suggest guidelines for these policies. Also suggested are ways for commanders to utilize offensive operations, intelligence, and rules of engagement to limit the degree of stress experienced by servicemen. In addition, commanders must see to servicemen's physiological needs by ensuring adequate amounts of sleep along with nutrition and water intake, just as they must be sure to maintain morale during military operations. Noncommissioned officers bear considerable responsibility in attaining these goals. Finally, following military operations, many military and civilian authorities advocate the use of psychological debriefings to prevent the subsequent development of PTSD. The authors review data that question the validity of this approach.

Acknowledgments

The authors would like to thank a number of individuals who contributed to this monograph. Special appreciation goes to Dr. Martin Iguchi, who facilitated the present collaboration. Several individuals were especially helpful in providing important suggestions and resources: Colonel Elspeth Ritchie (USA MC), Colonel Robert J. T. Joy (USA MC, ret.), Dr. Simon Wessely, Colonel Philip Volpe (USA MC), Lieutenant Colonel Tim Thomas (USA, ret.), Dr. John Greenwood, and Colonel Ron Levy (IDF, ret.). Scott Gerwehr provided many helpful comments on previous drafts. Thanks go to Dr. David H. Marlowe, COL James W. Stokes (USA MC), and Lieutenant General George R. Christmas (USMC, ret.) for their careful reviews and considered critiques. We are especially grateful for the many people interviewed as part of this research. While not all are cited, all contributed immensely to this report. All remaining weaknesses are the sole responsibility of the authors.

Acronyms

CED	Critical Event Debriefing
CISD	Critical Incident Stress Debriefing
CSC	Combat Stress Control
CSR	Combat Stress Reaction
CSS	Combat Service Support
DEROS	Date of Expected Return from Overseas
DNBI	Disease and Non-Battle Injury
FM	Field Manual
ID	Infantry Division
IDF	Israeli Defense Forces
KIA	Killed in Action
MC	Medical Corps
MH	Mental Health
MOOTW	Military Operations Other Than War
MOS	Military Occupational Specialty
MRE	Meals Ready to Eat
NCO	Noncommissioned officer
OIF	Operation Iraqi Freedom

PD	Psychological Debriefing
PDF	Panamanian Defense Force
PIES	Proximity, Immediacy, Expectancy, and Simplicity
POW	Prisoner of War
PT	Physical Training
PTSD	Post-Traumatic Stress Disorder
RCT	Randomized Controlled Trial
ROE	Rules of Engagement
RTD	Return to Duty
SET	Stress Exposure Training
TFR	Task Force Ranger
TOE	Table of Organization and Equipment
USA	United States Army
USMC	United States Marine Corps
WIA	Wounded in Action
WRAIR	Walter Reed Army Institute of Research

CHAPTER ONE

Introduction

> The neurotic compromise, in these circumstances, consists in a breakdown of an otherwise normal individual's ability to deal with his mounting anxiety and hostility in an efficient manner. Fear and anger in small doses are stimulating and alert the ego, increasing its efficiency. But when stimulated by repeated psychological traumata, the intensity of the emotion heightens until a point is reached at which the ego loses its effectiveness
>
> —*Roy Grinker and John Spiegel*[1]

Combat is stressful. Beyond the ever-present fear of death, servicemen and women of all branches must deal with an array of evolving emotional and physical environments that have the potential to exact a high psychological price. The boredom that comes with awaiting the next operation mixes with fear of the dire events that may occur. Across the line of departure arrives fear of the bullet that bears one's name, frustration of opportunities missed, joy that comes with victory, and immeasurable grief for those who have fallen. Carnage is ever present, and the fight to save the wounded can be even more daunting than the one to destroy the enemy. Physical exhaustion comes quickly as the body exerts the energy required to save one's

[1] R. R. Grinker and J. P. Spiegel, *Men Under Stress* (New York: McGraw Paperbacks, 1963), p. 78.

own life and the lives of comrades while simultaneously taking the lives of the enemy. Sleep loss, malnutrition, and dehydration are ever-present. Also common are the physical elements (heat/cold, rain/drought) that may reach extremes. All of these factors combine to weaken the resolve of mind and body.

This combat-induced stress can be a commander's ally or foe. Stress triggers the sympathetic nervous system to steer blood to the heart, brain, and skeletal muscles along with the release of hormones that increase heart rate and blood sugar.[2] These physiological reactions prepare the body for increased vigilance, physical stamina, speed, aggression, and pain tolerance. In addition, the stress of combat solidifies the bonds of friendship and loyalty among fellow combatants. It can trigger great acts of selflessness and heroism.

Unfortunately, stress has a dark side. The stress of battle may engender an array of dysfunctional behaviors from misconduct to psychiatric reactions. Misconduct behaviors vary from purposefully killing noncombatants, such as civilians or prisoners of war (POWs), to drug use, looting, disobedience, self-inflicted wounds, or desertion. While these problem behaviors are more common in low-intensity conflicts (LIC),[3] they also happen in more operationally intense theaters. An example of the latter was observed by E. G. Sledge, who witnessed a fellow marine at the Battle of Peleliu drive his Ka-Bar knife into the mouth of a living Japanese prisoner with the simple goal of extracting a gold tooth.[4]

Psychiatric reactions comprise the other type of response to stress. Post-traumatic stress disorder or PTSD is the most well known of this type of reaction. PTSD is a psychiatric disorder triggered by exposure to an extreme traumatic stressor such as combat or the witnessing of dead and dying persons. The symptoms this disorder engenders are chronic, lasting at least a month but often continuing for

---

[2] In contrast, the parasympathetic nervous system is associated with the body's relaxed state.

[3] S. Noy, "Battle Intensity and the Length of Stay on the Battlefield as Determinants of the Type of Evacuation," *Military Medicine,* Vol. 152, No. 12 (1987), pp. 601–607.

[4] E. B. Sledge, *With the Old Breed at Peleliu and Okinawa* (New York: Oxford University Press, 1981).

years. Symptoms include continual re-experiencing of the trauma, avoidance of events or people associated with it, and a lack of interest in realms unrelated to the trauma. Sleep problems, irritability, or an exaggerated startle response also plague such individuals.[5] Prevalence rates of PTSD vary between conflicts. Almost 20 years after the conclusion of Vietnam, 15.2 percent of surveyed veterans met criteria for PTSD.[6] In contrast, only 1.2 percent of Persian Gulf War veterans met diagnostic criteria.[7] Despite such variation, it is clear that clinical PTSD afflicts only a minority of combat veterans.

One common psychiatric response to combat is the combat stress reaction (CSR) or battle fatigue. Also known as neuropsychiatric casualty, combat exhaustion, and battle shock,[8] CSR is defined as any response to battle stress that renders a soldier combat ineffective. Symptoms of CSR vary widely, from debilitating anxiety or depression to hallucinations or even paralysis, from freezing under fire to outright panic.[9]

CSR can have a significant impact on an army's ability to wage war. In World War II, combat stress casualties ranged from 20 to 30 percent of those wounded in action (WIA).[10] In some battles, such as Okinawa, stress casualties are reported to have reached one for every

---

[5] American Psychiatric Association, *Diagnostic and Statistical Manual for Mental Disorders*, 4th ed. (Washington, D.C.: American Psychiatric Association, 1994).

[6] R. A. Kulka et al., *Trauma and the Vietnam War Generation* (New York: Brunner/Mazel, 1990).

[7] The Iowa Persian Gulf Study Group, "Self-Reported Illness and Health Status Among Gulf War Veterans: A Population-Based Study," *Journal of the American Medical Association,* Vol. 277, No. 3 (1997), pp. 238–245.

[8] While these different names are reflective of the eras in which they were coined, we will use all of them interchangeably throughout this report.

[9] Department of Defense, *Leaders' Manual for Combat Stress Control* (Department of the Army, FM 22-51), http://www.vnh.org/FM22-51/booklet1.html (last accessed May 22, 2004).

[10] L. J. Thompson et al., "Neuropsychiatry at Army and Division Levels," in W. S. Mullins and A. J. Glass (eds.), *Medical Department, United States Army Neuropsychiatry in World War II*, Vol. 2, *Overseas Theater* (Washington, D.C.: U.S. Government Printing Office, 1973), pp. 275–374.

two physical casualties.[11] Even in more recent armed conflicts such as the Israeli invasion of Lebanon (1982), stress casualties averaged about 23 percent of those wounded, with certain units bearing an even greater load, having nearly equal rates of CSR and WIA.[12] It is no doubt that such a loss of combat manpower poses a considerable threat to the lives and welfare of all troops involved. At a personal level, combat stress reactions may also increase the risk for PTSD.[13]

Critically, commanders have at their disposal the means to limit the numbers afflicted with acute stress reactions. Evidence of this comes from World War II and other conflicts in which were seen high interunit variability in stress casualty rates. In brief, among units that were highly trained, well led, and marked by high levels of morale and unit cohesion, stress casualties amounted to just 5 to 10 percent of their wounded. Moreover, all the steps that work to reduce stress simultaneously improve combat effectiveness. Thus, the actions of a given command influence the extent to which their soldiers develop acute stress reactions and, in turn, their level of combat effectiveness.

Also important are the actions taken by a unit once stress reactions arise. If the symptoms of CSR are known, as are the specific conditions under which it develops, early symptomatic soldiers can be identified. Once identified, simple action steps, taken within the unit, may help stem the reaction's tide and return the soldier to normal functioning. In more severe and/or unremitting cases, mental health (MH) personnel organic to the divisions and brigades, and/or in mobile teams from combat stress control (CSC) units,[14] may be available

---

[11] Department of Defense, *Leaders' Manual for Combat Stress Control.*

[12] COL Gregory Belenky, M.D., interview with the author, Washington, D.C., May 20, 2003.

[13] Z. Solomon, R. Benbenishty, and M. Mikulincer, "A Follow-Up of Israeli Casualties of Combat Stress Reaction ('Battle Shock') in the 1982 Lebanon War," *British Journal of Clinical Psychology,* Vol. 27, Pt. 2 (1988), pp. 125–135.

[14] CSC units are medical support units that specialize in the prevention and treatment of mental health–related problems. During combat operations these units constitute a second treatment echelon, along with division and possibly brigade mental health assets, for soldiers afflicted with CSR.

to provide treatment. Early and appropriate treatment results in enhanced return-to-duty (RTD) rates. Leadership has a significant impact on the extent and severity of combat stress reactions within a unit. Given the risks posed by combat stress reactions to both the safety of the tactical unit and the well-being of the individual soldier, officers and NCOs alike must develop a clear understanding of the nature of CSR and its treatment and prevention. This monograph thus seeks to give leaders a review of individual and environmental factors known to precipitate stress reactions, an overview of its battlefield treatment, and a discussion of preventive measures.

The Urban Dilemma

> There is no species of duty in which the soldier is liable to be employed so galling or so disagreeable as a siege . . . . Not that it is deficient in causes of excitement, which, on the contrary, are in hourly operation; but it ties him so completely down to the spot, and breaks in so repeatedly upon his hours of rest, and exposes him so constantly to danger, and that too at times and places where no honour is to be gained, that we cannot greatly wonder at the feelings of absolute hatred which generally prevail, among the privates at least of a besieging army, against the garrison which does its duty to its country by holding out to the last extremity.
>
> —*George Gleig* [15]

While all combat is stressful, urban operations entail a number of distinctive features that may place unique psychological demands upon combatants. The urban environment is highly threatening. The three-

[15] G. R. Glieg, *The Subaltern: A Chronicle of the Peninsular War* (Edinburgh, 1877), referenced in R. Holmes, *Redcoat: The British Soldier in the Age of Horse and Musket* (London: HarperCollins Publishers, 2001), pp. 378–379.

dimensional space from which defenders can engage friendly forces results in innumerable fields of fire and a dearth of concealment positions for forces that must often attack at ground level. Most friendly forces operate within a 50-meter radius of enemy soldiers who must often be engaged at even closer quarters during room-to-room clearances in buildings. Due to the presence of civilians, rules of engagement (ROE) may require visual confirmation of enemy targets and limit the use of indirect fire or aerial bombardment. Tall buildings may block artillery fire or limit radio communication between operating units and command.

Given these features, military doctrine for centuries has counseled against waging war in cities. Sun Tzu, for example, argued that "the worst policy is to attack cities. Attack cities only when there is no alternative."[16] The doctrinal policy of the Cold War–era Soviet military made similar statements.[17] Unfortunately, population trends show that greater numbers of people are flocking to live in urban areas.[18] In addition, as the Gulf War demonstrated, U.S. forces are nearly unmatched in terms of combat on open terrain. Fighting in densely packed city streets, however, limits America's vastly superior firepower, pitting rifleman against rifleman. For these reasons, U.S. and allied forces will inevitably be charged to operate militarily in the city's confines. Their only recourse is to be prepared.

Accordingly, in addition to providing a general review of the treatment and prevention of stress reactions, this monograph pays special attention to urban-specific implications of CSR. Central to this review is a study of the psychological risks posed by urban operations. Accurately evaluating the risk of CSR in urban operations is critical. It is necessary for the approximation of combat troop

---

[16] S. Tzu, *The Art of War,* Samuel B. Griffith (trans.) (New York: Oxford University Press, 1982), p. 78.

[17] R. W. Glenn, *Combat in Hell: A Consideration of Constrained Urban Warfare* (Santa Monica, CA: RAND Corporation, MR-780-A/DARPA, 1996).

[18] In 1950 and 1998, 22 percent and 50 percent respectively of the world's population lived in large urban centers. It is estimated that in 2010, 75 percent of the population will live in large urban centers. D. M. L. Chupick, "Training for Urban Operations," *Dispatches: Lessons Learned for Soldiers,* Vol. 9, No. 2 (2002), pp. 3–42.

strength. Physical casualty rates are typically high in this environment, and unforeseen depletion in combatant manpower can pose a significant danger to remaining forces. Similarly, estimation of stress casualties is necessary so that both combat and medical support units will be able to anticipate and consequently plan for the type and severity of stress reactions.

CHAPTER TWO

A Look Back: A Brief History of Combat Psychiatry

Events of our past not only remind us of our weaknesses and fallibilities but also serve to identify future directions. This is the case with the history of combat, or forward, psychiatry. Referring to psychiatry conducted at the "forward" lines, forward psychiatry seeks to return soldiers afflicted with stress reactions to their units and to limit the numbers of stress-related out-of-theater evacuations. Humans by their very nature are vulnerable. Many soldiers subjected to the horrors of battle experience a psychological breakdown. Military commanders should understand that their soldiers and marines will always be at risk. However, the risk is not a blind one. An education of the past helps clarify which steps have worked and which have not. It also clarifies the conditions that pose the greatest risk for stress casualties. Finally, history serves as a reminder of the importance of heeding prior lessons learned.

U.S. Civil War and Before

Though battle fatigue or CSR was not recognized in its 20th century form, soldiers who fought in the U.S. Civil War did not escape the psychological consequences of battle. Desertion was prominent in both armies, and it was not uncommon for soldiers to panic in the midst of battle.[1] Soldiers also suffered from "nostalgia" and "soldier's heart." Nostalgia was a malady whose name was first coined by a Swiss doctor in 1678. The disorder was characterized by feelings of homesickness and explosive aggression, disciplinary problems, social estrangement, constricted affect, and mistrust of command, and it generally occurred when soldiers were encamped away from home for long periods.[2] Symptoms generally dissipated when soldiers prepared for battle.[3] In contrast, soldier's heart was characterized by an ambiguous cardiac condition consisting of an increased heart rate and fatigue with no discernable medical problems present. It was a problem that would presage the "disorders of the heart" syndrome commonly found in the Boer War and World War I.[4]

The Great War

> Are you in the full glory of manly strength?
> Are you a man in every sense of the word?
>
> —*A 1915 advertisement*[5]

World War I began as a war of movement, during which few acute psychological reactions to battle were observed. The battle lines eventually stabilized and trench warfare came to the fore. Suddenly, sol-

---

[1] D. H. Marlowe, *Psychological and Psychosocial Consequences of Combat and Deployment with Special Emphasis on the Gulf War* (Santa Monica, CA: RAND Corporation, MR-1018/11-OSD, 2001).

[2] F. D. Jones, "Psychiatric Lessons of War," in F. D. Jones et al. (eds.), *War Psychiatry* (Washington, D.C.: TMM Publications, 1995a), pp. 1–33.

[3] Marlowe, *Psychological and Psychosocial Consequences.*

[4] E. Jones et al., "Post-Combat Syndromes from the Boer War to the Gulf War: A Cluster Analysis of Their Nature and Attribution," *BMJ (British Medical Journal),* Vol. 324 (2002), pp. 321–324.

[5] *Daily Mirror,* July 12, 1915, quoted in B. Shephard, *A War of Nerves* (London: Jonathan Cape, 2000), p. 15.

diers presented dramatic symptoms that included paralysis, blindness, and amnesia, along with more subtle symptoms such as headache, sleeplessness, depression, and anxiety.[6] The British approach to mental illness limited the extent to which fear could account for these symptoms, as it was believed that only women and not men suffered from psychological disorders.[7] These battle-induced symptoms must be something other than mental illness. Soldiers and medical authorities alike focused their attention on the high-explosive artillery shells rained down by the Germans. The combustive effects of these shells, it was presumed, sent a shock wave at a soldier's head, causing microhemorrhages of the brain. They called it "shell shock." Eventually, however, authorities realized that soldiers did not have to be near an exploding shell to develop symptoms akin to shell shock. Consequently, the diagnostic term was changed to "war neurosis."[8]

The dramatic nature of shell shock or war neurosis symptoms was probably influenced by the expectations of command, medical authorities, and society in general. The prevailing culture held a masculine view of men. Expressions of fear were not encouraged and cowardice was unacceptable. Male mental illness was frowned upon. Soldiers who could no longer tolerate the strains of war had no other means of unconsciously communicating their inability to continue than to manifest symptoms suggestive of a physical disorder.[9] Oftentimes, soldiers who presented only psychological symptoms were treated in a disciplinary manner or, in a number of cases, sentenced to execution.[10]

---

[6] F. D. Jones, "Traditional Warfare Combat Stress Casualties," in F. D. Jones et al. (eds.), *War Psychiatry* (Washington, D.C.: TMM Publications, 1995b), pp. 35–61.

[7] B. Shephard, "Shell-Shock on the Somme," *RUSI Journal* (June 1996), pp. 51–56.

[8] A. Babington, *Shell-Shock: A History of the Changing Attitudes to War Neurosis,* (London: Leo Cooper, 1997). According to this same source, N.Y.D.N. (Not Yet Diagnosed Nervous) was also applied as a diagnostic term. In late 1915, the British army directed that true "shell-concussion" cases, those a result of enemy action, were deserving of a wound stripe. To determine this, physicians frequently held the patient in the casualty clearing stations while his unit was contacted. Until decided, N.Y.D.N. denoted the patient's status.

[9] Marlowe, *Psychological and Psychosocial Consequences*.

[10] Shephard, *A War of Nerves*, p. 223.

The treatment of war neurosis and the subsequent outcomes of those treatments evolved steadily throughout the war. Initially, afflicted soldiers were evacuated from the trenches to rearward base hospitals or to England. The distance from the line of contact weakened the soldier's sense of duty to his comrades. Coupled with the expectation of evacuation, symptoms of war neurosis were reinforced and ingrained in the wounded soldier's mind. Many of these soldiers were lost to the army for good. Those who remained in Great Britain often demonstrated a longstanding disability.

In 1916, Lieutenant Colonel Charles Myers, a psychologist with the British Army, sought to treat soldiers with war neurosis closer to the trenches. After observing the forward treatments administered by French neurologists, Dr. Myers created four separate treatment centers close to the front line. Soldiers were fed, rested, and submitted to graduated exercise and military marches. Some psychotherapy was performed. RTD rates increased dramatically as a consequence of this forward treatment.[11]

By 1917 the United States was preparing to enter the war. The Army sent Thomas Salmon to Europe to determine the proper treatment for stress casualties. Salmon observed the centers created by Myers and French neurologists, recommending that American stress casualties be treated close to the front shortly after evacuation and with the expectation that they return to duty. This treatment was subsequently described as "proximity, immediacy, and expectancy" (PIE) by Kenneth Artiss.[12] The term "simplicity" (food, drink, sleep) was later added, which expanded the acronym to PIES.

Dr. Salmon also established the multi-echelon treatment system whereby war neurosis patients were treated at locations with varying distances to the front. Stress casualties were first seen in advanced field hospitals. Those patients not returned to their units within a few days were then sent on to divisional hospitals where treatments were

---

[11] E. Jones and S. Wessely, "Psychiatric Battle Casualties: An Intra- and Interwar Comparison," *British Journal of Psychiatry,* Vol. 178 (2001), pp. 242–247.

[12] K. L. Artiss, "Human Behaviour Under Stress: From Combat to Social Psychiatry," *Military Medicine,* Vol. 128 (1963), pp. 1011–1015.

overseen by newly appointed division psychiatrists.[13] Return-to-duty rates for the advanced field hospitals were reported to reach as high as 80 percent, while the divisional hospitals returned approximately 65 percent of patients. Beyond division, stress casualties were treated at a group of neurological hospitals and even further from the front at a hospital dedicated to the treatment of war neurosis.[14]

World War II

As World War II approached, U.S. forces were determined to limit psychiatric losses of the scale suffered during World War I and the subsequent $42 million paid to psychiatrically afflicted veterans in 1940 alone.[15] But instead of heeding the lessons of World War I forward psychiatry, the U.S. military sought to limit psychiatric casualties by preventing psychologically vulnerable individuals from entering the service.

Potential recruits were rejected from service for educational deficiencies, assertions of anxiety disorders, or neurotic personalities.[16] Overall, 1,600,000 recruits were denied military service on the grounds of psychological or educational deficiencies.[17] More servicemen were rejected than were deployed overseas at certain junctures of the war.[18]

The screening program's success was tested in the war's early phases. U.S. marines at the Battle of Guadalcanal in August 1942

---

[13] Division psychiatrists were also tasked with forward preventive and triage missions and helped train division rear medical personnel in the treatment of stress casualties. COL James Stokes, M.D., written comments to the author, November 18, 2004.

[14] Babington, *Shell-Shock: A History of the Changing Attitudes to War Neurosis.*

[15] Shephard, *A War of Nerves.*

[16] Marlowe, *Psychological and Psychosocial Consequences.*

[17] A. J. Glass, "Lessons Learned," in A. J. Glass and R. J. Bernucci (eds.), *Medical Department, United States Army Neuropsychiatry in World War II,* Vol. 1, *Zone of Interior* (Washington, D.C.: U.S. Government Printing Office, 1966), pp. 735–760.

[18] F. D. Jones, "Psychiatric Lessons of War."

fought a four-month pitched battle. They "endured poor food, tropical diseases, sleeplessness and unceasing attack from land and air."[19] Psychiatric casualties were notoriously high, overwhelming the rearward treatment hospitals. Most were evacuated to Australia, New Zealand, and the United States.[20] A similar crisis of manpower occurred during the first American battles in the North African desert at the Kasserine and Faid Passes against German forces in early 1943. Poorly trained U.S. forces confronted the formidable Erwin Rommel's Afrika Corps. To make matters worse, soldiers in the heat of battle watched helplessly as American-made 75mm howitzer rounds bounced off the enemy's tanks.[21] The combined psychological impact of these and other factors was tremendous. The rates of psychoneurosis, the name given to stress reactions at the time, were in some units equivalent to the numbers of those wounded in action. The clinical picture was reminiscent of the dramatic conversion states[22] first observed in the shell-shocked victims of World War I. Similarly reminiscent were the hospitals far to the rear in Algeria, from which relatively few soldiers were returned to duty.[23]

Help came in the form of a neurologist named Frederick R. Hanson. Dr. Hansen and his colleague, Dr. Louis L. Tureen, conducted two demonstrations that reignited the principles of forward psychiatry. In the first, they reportedly returned to combat 30 percent of patients in a corps clearing station within 30 hours of arrival, and in a second, they returned 70 percent to combat and most of the remainder to base section duty.[24] In April 1943, Hanson even peti-

---

[19] Shephard, *A War of Nerves,* p. 223.

[20] Shephard, *A War of Nerves.*

[21] Marlowe, *Psychological and Psychosocial Consequences.*

[22] Conversion states involve the expression of emotional conflicts as physical symptoms, e.g., extreme displeasure of combat is expressed through paralysis or severe back pain.

[23] C. S. Drayer and A. J. Glass, "Introduction," in A. J. Glass (ed.), *Medical Department, United States Army, Neuropsychiatry in World War II,* Vol. 2, *Overseas Theaters* (Washington D.C.: Office of the Surgeon General, Department of the Army, 1973).

[24] Drayer and Glass, "Introduction," p. 9. For a critical discussion of RTD rates, see Chapter Five, specifically the section entitled "The Success of PIES."

tioned General Omar N. Bradley to use a diagnostic term less suggestive of illness than psychoneurosis. Bradley consequently ordered that new cases of breakdown be labeled "exhaustion," the implication being that psychiatric breakdown was the result of natural fatigue and that simple rest was sufficient to return men to duty.[25] The subsequent symptoms of stress reactions bore this expectancy out, as soldiers generally appeared with such symptoms as restlessness, irritability, insomnia, depression, and anxiety.

Despite the psychological losses incurred in the Guadalcanal and North Africa campaigns, it would still take months before the U.S. military took full measures to treat and prevent combat exhaustion. One important stimulus was the well-publicized berating of a psychiatric casualty by Lieutenant General George S. Patton during a hospital visit near Palermo in August 1943.[26] The event caused a public uproar. Patton was forced to apologize, and his actions inadvertently brought the issue of combat exhaustion to the fore. In the subsequent invasion of Italy, plans were finally made to adopt the principles of forward psychiatry envisioned by Thomas Salmon.[27] Still, it took an additional five months until the War Department authorized the addition of a psychiatrist to U.S. divisions' Table of Organization and Equipment (TOE). These division psychiatrists generally acted as a first echelon treatment provider, with 300-cot "neuropsychiatric centers" serving as a second echelon in each corps area.[28]

Actions such as this were necessary given the high rates of combat exhaustion that had existed throughout the war. Overall rates of stress casualties to WIA varied from roughly 5 to 30 percent, as shown in Table 2.1.

---

[25] Drayer and Glass, "Introduction," pp. 9–10. E. Jones and S. Wessely, "Forward Psychiatry in the Military: Its Origins and Effectiveness," *Journal of Traumatic Stress,* Vol. 16, No. 4 (August 2003), pp. 411–419. The term "exhaustion" subsequently evolved during the war to "combat exhaustion" and "combat fatigue."

[26] It is known that Patton berated two soldiers near Palermo; however, one of them purportedly suffered from malaria, not combat exhaustion.

[27] Drayer and Glass, "Introduction."

[28] Stokes, written comments.

Table 2.1
Rates of WIA, Psychoneurosis, and Corresponding Percentage of Psychoneurosis as a Function of WIA for a Subset of Seventh U.S. Army Divisions, January 1 to May 15, 1945

Division	Total WIA	Total Psychiatric	Percentage of Psychiatric
12th Armored	5,202	1,011	19.4
14th Armored[a]	2,536	322	12.7
3rd Infantry	6,955	527	7.5
36th Infantry	3,662	785	21.4
42nd Infantry[b]	2,243	345	15.4
44th Infantry	2,438	688	28.2
45th Infantry	4,923	1,024	20.8
100th Infantry	3,104	822	26.5
103rd Infantry	2,555	820	32.1

[a] January 1–April 15, 1945.

[b] February 1–May 15, 1945.

SOURCE: Adapted from L. J. Thompson et al., "Neuropsychiatry at Army and Division Levels," p. 363.

Post–World War II

Korea

When North Korea invaded the south, many of the soldiers mobilized by the United States as part of Task Force Smith and other units were pulled from the Japanese occupation force. These forces were poorly equipped and trained.[29] Exhaustion casualties were high (250 per thousand per year). Given the absence of forward psychiatric treatment centers, many were evacuated to Japan or the United States.[30]

---

[29] Robert H. Mosebar, M.D., interview with the author, San Antonio, Texas, March 13, 2003.

[30] E. C. Ritchie, "Psychiatry in the Korean War: Perils, PIES, and Prisoners of War," *Military Medicine,* Vol. 167, No. 11 (2002), pp. 898–903.

The psychiatrist Colonel Albert Glass was sent to Korea as a consultant to establish forward psychiatry. Under his guidance, a three-echelon system of care was established in which the division psychiatrist constituted the first echelon, theater-level hospitals the second, and Japan- or U.S.-based hospitals the third. The division psychiatrist, now partnered with a psychologist, social worker, and enlisted personnel, also helped educate division surgeons and line officers in preventive psychiatry.[31] Dr. Glass created mobile mental health units called Medical Detachment, Psychiatric, or Team KO, which helped reinforce division psychiatrists during periods of heavy fighting.[32] These changes helped to reduce the tide of psychiatric casualties and increase combat fatigue RTD rates.

Vietnam

The Vietnam War can be construed as a conflict fought in three phases: an advisory period followed by full-scale combat operations and concluding with a slow-paced military withdrawal. The period of major combat operations was unique in that the rates of battle fatigue did not fluctuate with the peaks and troughs of physical casualties. Additionally, these rates (presented in Table 2.2) are markedly lower than those of World Wars I and II and the Korean War. There are several possible explanations for the differences. First, there was a 12-month rotation policy for enlisted servicemen and a 6-month command tour for officers. While booby traps were a constant concern, soldiers and marines were less regularly subjected to indirect fire and enemy engagements. When such engagements did occur, they could be extremely intense but were seldom other than relatively brief in duration. Wounded soldiers could also rely on helicopters to remove them safely and in a timely fashion from most any scene of battle. Regular rotation back into base camps further reduced exposure to the risks of "front-line" combat and likely played a role.

---

[31] Ritchie, "Psychiatry in the Korean War: Perils, PIES, and Prisoners of War." A. J. Glass, "Psychiatry in the Korean Campaign," *United States Army Medical Bulletin,* Vol. 4, No. 10 (1953), pp. 1387–1401. Stokes, written comments.

[32] Jones and Wessely, "Forward Psychiatry in the Military."

Table 2.2
Selected Causes of Admission to Hospital and Quarters Among Active Duty U.S. Army Personnel in Vietnam, 1965–1970

	Rate Expressed as Number of Admissions (per 1,000 Average Strength)					
	1965	1966	1967	1968	1969	1970
Wounded in action	61.6	74.8	84.1	120.4	87.6	52.9
Neuropsychiatric conditions	11.7	12.3	10.5	13.3	15.5	25.1
Venereal disease (includes CRO)	277.4	281.5	240.5	195.8	199.5	222.9

NOTE: "CRO" is carded for record only, i.e., not hospitalized.
SOURCE: Adapted from S. Neel, *Vietnam Studies: Medical Support of the U.S. Army in Vietnam, 1965–1970,* (Washington, D.C.: U.S. Department of the Army, 1973), referenced in F. D. Jones, "Traditional Warfare Combat Stress Casualties," p. 36.

While rotation may have reduced the number of psychiatric casualties in Vietnam, it is often credited with creating problems in unit cohesion and combat effectiveness. Every serviceman in theater knew the exact date of his DEROS (Date of Expected Return from Overseas). The realization that one did not have to get killed or wounded to return stateside, as was the expectation in World War II, raised hopes and individual morale. Unfortunately, with individual rotations, men more frequently fluctuated in and out of units, limiting the extent to which bonds of cohesion developed and solidified. In addition, abbreviated command tours sometimes resulted in stints of poor leadership, precipitating concomitant fears in enlisted personnel that they would pay the price in blood.[33] "Short-timer's syndrome" was also a problem. The approach of DEROS found many service members losing morale and combat effectiveness. It not un-

[33] These fears were born in truth. Soldiers under battalion commanders with more than six months experience averaged 1.62 killed in action per month per battalion, whereas soldiers under battalions with commanders with less than six months experience averaged 2.46 killed in action per month. T. C. Thayer (ed.), *A Systems Analysis View of the Vietnam War* (Alexandria, VA: Defense Technical Information Center, 1978).

commonly afflicted men to such a degree that they had to be removed from combat duties altogether.[34]

As in previous wars, a multi-echelon system of psychiatric care was developed. Soldiers were first treated in battalion aid stations, and those with persistent symptoms were airlifted[35] back to division base camp hospitals. The third echelon consisted of two neuropsychiatric specialty teams, which could hold and treat patients for up to 30 days before out-of-theater evacuations were necessitated.[36]

Combat operations were limited during the withdrawal phase of the war. While the rates of killed and wounded in action decreased, the rates of misconduct-type stress casualties began to rise. Substance abuse was by far the most common type of misconduct disorder, as illicit drug use rates in Vietnam climbed from 50 percent of U.S. servicemen in 1968 to 70 percent in 1973.[37] Servicemen committed other offenses as well, from refusing to obey orders to committing acts of violence against civilians and officers (fragging).[38] Such problems were especially common in support units. Evacuation out of theater for substance abuse and other character problems was not permitted in the U.S. Army until 1971. But once this policy changed, evacuations for psychiatric problems, including substance abuse, rose dramatically.[39] Subsequently, CSC doctrine adopted the concept of "misconduct stress behaviors" and leadership actions for preventing them were emphasized.[40]

---

[34] Marlowe, *Psychological and Psychosocial Consequences*.

[35] For those evacuated to separate treatment sites, helicopters may have compensated for less geographical proximity by facilitating their return to their units. Stokes, written comments.

[36] H. S. Block, "Army Clinical Psychiatry in the Combat Zone: 1967–1968," *American Journal of Psychiatry*, Vol. 126, No. 3 (1969), pp. 289–298.

[37] B. C. Dubberly, "Drugs and Drug Use," in S. C. Tucker (ed.), *Encyclopedia of the Vietnam War: A Political, Social and Military History* (Santa Barbara: ABC-CLIO, 1998), pp. 179, 180.

[38] J. McCallum, "Medicine, Military," in *Encyclopedia of the Vietnam War*, pp. 423–428.

[39] F. D. Jones, "Traditional Warfare Combat Stress Casualties."

[40] Stokes, written comments.

The most lasting contribution of Vietnam to the history of battle trauma is the legacy of post-traumatic stress disorder (PTSD). Shortly after the war, many veterans began reporting a variety of psychological problems such as apathy, depression, mistrust, insomnia, and nightmares, problems that were quickly attributed to their combat experience. What began unofficially as the "Post-Vietnam Syndrome" in the early 1970s became PTSD by 1980.[41] Since that time, the government has contributed vast financial resources to PTSD treatment in the form of research and treatment grants. Despite such resources, treatment has proven only moderately effective. Thousands continue to manifest the disorder's array of symptoms.[42]

The Israeli Wars

In 1973, Egypt launched a surprise attack on Israel. In this four-week battle, Israel was initially caught off guard and forced to retreat. Reserves were mobilized, and Israel gained its initial position and the tactical advantage. It was a high-intensity battle marked by continuous combined arms operations across a highly mobile battlefield. The combination of surprise and high-intensity warfare resulted in a ratio of stress casualty to wounded of 30:100. The suddenness with which these casualties were incurred caused Israel to forgo the term battle fatigue and instead opt for battle shock. Regardless of the name, Israel was ill prepared to handle these casualties. No doctrine for forward psychiatry existed, and most casualties were invalided back to Israel.[43]

Nine years later, Israel fought again, this time in Lebanon. The war was launched by the Israelis, who mobilized only a portion of their military. It was fought in both mountainous and urban (Beirut, Sidon, Tyre, and elsewhere) terrain. Though the ratio of "battle shock" casualties to WIA reached 23:100, Israel was more prepared to handle them than it was in the 1973 Yom Kippur War, as various

---

[41] Shephard, *A War of Nerves*.

[42] Kulka et al., *Trauma and the Vietnam War Generation*.

[43] G. L. Belenky, C. F. Tyner, and F. J. Sodetz, *Israeli Battle Shock Casualties: 1973 and 1982* (Washington, D.C.: Walter Reed Army Institute of Research, 1983).

treatment echelons were available at the war's start.[44] Israel had spent the intervening years developing a doctrine of forward psychiatry based on that of the U.S. Army. A significant percentage of psychological casualties were inadvertently sent far behind the lines of battle, but many were successfully treated in forward restoration centers.[45] Also, one new phenomenon was the occurrence of stress reactions after the battle's conclusion. In fact, 40 percent of all stress reactions occurred in the full year following hostilities.[46]

Developments and Experiences Since 1983

Following the war in Lebanon, U.S. Army medical authorities were stimulated to transform their forward psychiatric capabilities. Consequently, in 1985, according to Dr. James Stokes, "combat stress control" was recognized by the Army "as an autonomous Medical Department functional area" and "CSC became a distinct Battlefield Operating System." With this change came a dramatic increase in CSC-allocated TOE positions.[47]

For Operation Desert Shield, the U.S. Army deployed mental health teams with brigades from the XVIII and VII Corps. In addition, three OM Teams, the 43-person predecessor to the present-day CSC unit, were deployed. One of these teams set up a holding treatment center adjacent to the combat support hospital, which helped reduce the escalating psychiatric evacuation rate. Other teams helped assess units' perceived readiness for combat and cohesion and provided training in combat stress control. During the air campaign, the

---

[44] Ibid.

[45] Z. Solomon and R. Benbenishty, "The Role of Proximity, Immediacy, and Expectancy in Frontline Treatment of Combat Stress Reaction Among Israelis in the Lebanon War," *American Journal of Psychiatry*, Vol. 143, No. 5 (1986), pp. 613–617.

[46] S. Noy, R. Levy, et al., "Mental Health Care in the Lebanon War, 1982," *Israel Journal of Medical Sciences*, Vol. 20, No. 4 (1984), pp. 360–363.

[47] Stokes, written comments.

OM teams sent teams forward to provide holding treatment near the Kuwait border.[48]

Following the Gulf War, CSC units or other mental health assets participated in a number of operational deployments. Some of these include Operation Restore Hope in Somalia, Operation Uphold Democracy in Haiti, and Operation Joint Endeavor in Bosnia. In addition to treating cases of acute stress reactions, mental health units also commonly provided outpatient treatment for soldiers with preexisting mental health problems, command consultations, unit surveys, and stress management classes.[49] Mental health assets are, at the time of this writing, serving in both Afghanistan and Iraq. The history of forward psychiatry in both of these theaters is not yet written and as such will not be addressed in this monograph.

---

[48] Stokes, written comments. M. E. M. Doyle, "Combat Stress Control Detachment: A Commander's Tool," *Military Review* (May–June, 2000), pp. 65–71.

[49] E. C. Ritchie and D. C. Ruck, "The 528th Combat Stress Control Unit in Somalia in Support of Operation Restore Hope," *Military Medicine,* Vol. 159, No. 5 (1994), pp. 372–376. B. L. Bacon and J. J. Staudenmeijer, "A Historical Overview of Combat Stress Control Units of the U.S. Army," *Military Medicine,* Vol. 168, No. 9 (2003), pp. 689–693.

CHAPTER THREE

The Lessons of War: The Causations of Battle Fatigue

War has taught us much about the precipitants of acute stress reactions or battle fatigue. What follows is a summary of these lessons. They are ordered by individual, unit, and battlefield factors. These lessons are derived mostly from World War II and the Israeli wars, from which most CSR-related observations have been collected. The names and at times the symptomatology of CSR have changed over time. Despite such changes, the frequent agreement in findings from these conflicts suggests that the basic principles underlying breakdown is similar across generations. It is not complete; reappraisals of prior wars and future conflicts are sure to add to and subtract from these lessons, especially as the nature of warfare continues to evolve. However, the building blocks as presented here are substantive enough to help commanders and senior NCOs develop expectations as to what specific events and conditions increase the risk of battle fatigue and in turn limit overall combat effectiveness. They also provide a base from which to take preventive actions.

Individual Factors

> One of our cultural myths has been that only weaklings break down psychologically [and that] strong men with the will to do so can keep going indefinitely.
>
> —*Beebe and Appel*[1]

[1] G. W. Beebe and J. W. Appel, *Variation in Psychological Tolerance to Ground Combat in World War II* (Washington, D.C.: National Academy of Sciences, 1958), p. 164.

Personality

The Yom Kippur War was the only conflict in which the relationship between personality and combat stress reactions was studied. In this instance, investigators could not document a significant relationship between personality and stress.[2] Alternately, personality characteristics have been associated with reactions to other stressful situations and to combat-related performance. For example, the Human Resources Research Organization (HumRRO), working for the U.S. Army, conducted a series of studies called the "Task Fighter."[3] In one component, men from three combat infantry divisions in Korea were classified as being "Fighters" or "Non-Fighters." In the other, investigators evaluated performance in a series of stress-related tasks (fire fighting, bayoneting dummies in a darkened room, and psychomotor tasks with and without electric shock). Both studies concurred that high-performing individuals were more financially experienced in their lives, made more money, and were more interested in masculine activities such as poker, cars, and body-contact sports. In addition, the Korean War "Fighters" scored high on personality characteristics such as leadership and extroversion (i.e., sociability). Interestingly, the findings of masculinity may not differ greatly from a more recent examination in which the Norwegian researcher Jar Eid demonstrated that Norwegian military personnel who rated low on a measure of "hardiness" were more likely to experience stress-related symptoms during survival school training.[4]

Nonmilitary Stressors

One critical factor related to operational stress is the extent to which soldiers worry about non-military-related stressors. During World

---

[2] S. Noy, "Stress and Personality as Factors in the Causation and Prognosis of Combat Reactions," in G. L. Belenky (ed.), *Contemporary Studies in Combat Psychiatry* (Westport, CT: Greenwood Press, 1987), pp. 21–30.

[3] Reviewed in P. Watson, *War on the Mind: The Military Uses and Abuses of Psychology* (New York: Basic Books, 1978); most of the original research took place in the 1950s.

[4] Charles A. Morgan, M.D., interview with the author, New Haven, Connecticut, March 27, 2003.

War II, soldiers who left spouses back home worried more about their families and about the likelihood of becoming a casualty than soldiers who were unmarried.[5] In the Yom Kippur War, such worries were directly tied to CSR. It was found that 80 percent of Israeli CSR casualties reported either prior or ongoing civilian stressors such as a pregnant spouse or birth of a new child in the last year (50 percent of the sample) or the recent death of a loved one (23 percent). Also, 40 percent of the sample reported a recent marriage, a new mortgage, sick parents, or other personal stressors.[6] Though marital status and major life events were not related to acute stress reactions in the Lebanon War,[7] they were related to the subsequent development of PTSD symptoms.[8] As the medical anthropologist David Marlowe states, "Almost all work done on stress has consistently demonstrated that stressors are additive and probably cumulative. New stressors do not displace old ones. The stresses of the deployment are added to the ones brought from or generated at home."[9]

Education

Level of education also appears related to the occurrence of stress reactions. In World War II, Stouffer demonstrated that relatively undereducated soldiers were more likely to report higher levels of anxiety than educated soldiers for both fresh and veteran replacements going overseas.[10] Education also appeared to be a factor in Israel's 1982 war in Lebanon, as 27 percent of stress casualties in comparison to only 12 percent of physical casualties had fewer than 8 years of

---

[5] S. A. Stouffer et al., *The American Soldier, Combat and Its Aftermath* (New York: John Wiley & Sons, Inc., 1965).

[6] Belenky, Tyner, and Sodetz, *Israeli Battle Shock Casualties.*

[7] Z. Solomon and H. Flum, "Life Events and Combat Stress Reaction in the 1982 War in Lebanon," *Israel Journal of Psychiatry and Related Sciences,* Vol. 23, No. 1 (1986), pp. 9–16.

[8] Z. Solomon and H. Flum, "Life Events, Combat Stress Reaction and Post-Traumatic Stress Disorder," *Social Science and Medicine,* Vol. 26, No. 3 (1988), pp. 319–325.

[9] Marlowe, *Psychological and Psychosocial Consequences*, p. 123.

[10] Stouffer et al., *The American Soldier*.

education.[11] Given the low levels of education reported in the preceding samples, it seems likely that they hold little relevance to the relatively highly educated members of today's U.S. military.

Unit Factors

Morale

> When fire sweeps the field, be it in Sinai, Pork Chop Hill or along the Normandy coast, nothing keeps a man from running except a sense of honor, of bound obligation to people right around him, of fear of failure in their sight which might eternally disgrace him.
>
> —*S. L. A. Marshall*[12]

Morale refers to the general sense of well-being enjoyed by the military unit. When morale has been evaluated in the broad sense of individual and unit well-being, it is clear from both the Yom Kippur and Lebanon wars that retrospectively recalled levels of morale were disproportionately low in soldiers with CSR.[13] According to Anthony Kellett, three factors that contribute to this well-being are self-confidence, trust in command, and unit cohesion.[14] Each of these factors is discussed further below.

---

[11] Z. Solomon, S. Noy, and R. Bar-On, "Risk Factors in Combat Stress Reaction: A Study of Israeli Soldiers in the 1982 Lebanon War," *Israel Journal of Psychiatry and Related Sciences*, Vol. 23, No. 1 (1986), pp. 3–8.

[12] S. L. A. Marshall, "Combat Leadership," in *Preventive and Social Psychiatry* (Washington, D.C.: U.S. Government Printing Office, 1957), p. 305.

[13] M. Steiner and M. Neumann, "Traumatic Neurosis and Social Support in the Yom Kippur War Returnees," *Military Medicine*, Vol. 143, No. 12 (1978), pp. 866–868, and Belenky, Tyner, and Sodetz, *Israeli Battle Shock Casualties*.

[14] A. Kellett, *Combat Motivation* (Boston, The Hague, London: Kluwer Nijhoff Publishing, 1982).

Self-Confidence

Servicemen who lack confidence in their military skills and weapons are at considerable risk for breakdown. For example, prior to the Normandy landing in World War II, heavy weapons companies in four separate divisions rated their willingness for combat, their confidence in combat stamina, and confidence in combat skills. Those companies that reported the lowest levels of confidence experienced a disease and nonbattle injury rate that was almost twice that of companies reporting high self-confidence levels.[15] Similarly, during the Yom Kippur War, almost half of reserve soldiers with stress reactions reported low self-esteem about professional military knowledge.[16]

Faith in Command

A lack of confidence in command has also been related to both stress levels and psychiatric breakdown. In the 1973 Israeli-Arab War, 42 percent of a group of reserve soldiers with either CSR or PTSD type symptoms reported feeling "no trust toward immediate command," compared with 5 percent of a group of unaffected elite soldiers.[17] More recently, the Walter Reed Army Institute of Research (WRAIR) study of Gulf War stress found that units with leadership problems prior to deployment tended to exhibit the most stress and had difficulties coping with that stress.[18]

Cohesion

Unit cohesion is, by most accounts, the most important factor in the prevention of psychiatric breakdown in battle and is one of the greatest stress-related lessons of World War II. In reference to that war, S.L.A. Marshall stated that "the unit was the primary defender against or expediter of breakdown in battle." The psychiatrist Herbert Spiegal made a similar but more eloquently stated observation:

---

[15] Stouffer et al., *The American Soldier*.

[16] Steiner and Neumann, "Traumatic Neurosis and Social Support."

[17] Ibid.

[18] Marlowe, *Psychological and Psychosocial Consequences*.

> If abstract ideas—hate or desire to kill—did not serve as strong motivating forces, then what did serve them in that critical time? What enables them to attack, and attack, and attack week after week in mud, rain, dust, and heat until the enemy was smashed? It seemed to me that the drive was more a positive than a negative one. It was love more than hate. Love manifested by 1) regard for their comrades who shared the same dangers, 2) respect for their platoon leader or company commander who led them wisely and backed them with everything at his command, 3) concern for their reputation with their commander and leaders, and 4) an urge to contribute to the task and success of their group and unit.
>
> In other words, interpersonal relationships among men and between the men and their officers became more intense and more important. These cohesive forces enabled them to identify themselves as part of their unit. It enables them to muster and maintain their courage in the most trying situations. It even led them at times to surprise themselves with gallant and heroic actions. They seemed to be fighting for somebody rather than against somebody.[19]

Data from the most recent Israeli wars substantiate these observations. In the Yom Kippur War, Shabtai Noy found that 40 percent of a group of CSR casualties reported mistrust in their commanders and felt socially isolated in their units.[20] Social isolation was also the "best single predictor of CSR" in a retrospective study of Israeli stress casualties from the Lebanon War.[21]

Combat Assignment

In World War II, the line infantryman or dismounted armored infantrymen succumbed in the greatest rates to the stressors of warfare.

---

[19] Marlowe, *Cohesion, Anticipated Breakdown, and Endurance in Battle: Considerations for Severe and High Intensity Combat* (Washington, D.C.: WRAIR, 1979), pp. 24, 25.

[20] Noy, "Stress and Personality."

[21] Z. Solomon et al., "Effects of Social Support and Battle Intensity on Loneliness and Breakdown During Combat," *Journal of Personality and Social Psychology*, Vol. 51, No. 6 (1986), pp. 1269–1276.

High psychoneurosis rates for World War II–era regimental combat units, in comparison to other units, are displayed in Table 3.1. During more recent conflicts, reserve soldiers and members of support units have reported psychiatric casualty rates disproportionately greater than would be expected given their rates of WIA.[22] In the Yom Kippur War, the ratio of psychiatric to physical casualties was three times greater for support units than for combat (3.0 versus 0.8).[23] During the air campaign of Operation Desert Storm, support units constituted 81 percent of a sample of stress and psychiatric casualties from the U.S. Army's 7th Corps.[24] In the case of reserve forces, 80 percent of all known stress casualties from the Lebanon War were members of reserve forces, compared with 46 percent of those who were WIA.[25]

There are a number of potential reasons for high stress rates among both reserve and support units. Low levels of cohesion is one

Table 3.1
Rate of U.S. Army Psychoneurosis Admissions as a Function of Military Occupational Specialty for Those Deployed Overseas in 1944

Military Occupational Specialty	Mean Strength	Number of Psychoneurosis Admissions	Rate per 1,000 per Annum	As Percentage of Mean
Regimental combat	214,000	61,150	285	28.5%
Other units	2,004,600	41,950	13	1.3%
Total	3,259,200	102,100	31	3.1%

SOURCE: Adapted from N. Q. Brill, G. W. Beebe, and R. L. Lowenstein, "Age and Resistance to Military Stress," p. 1260.

---

[22] It is not known whether the ratio of stress casualties to WIA was greater for support units during World War II.

[23] I. Levav, H. Greenfield, and E. Baruch, "Psychiatric Combat Reactions During the Yom Kippur War," *American Journal of Psychiatry*, Vol. 136, No. 5 (1979), pp. 637–641.

[24] D. R. McDuff and J. L. Johnson, "Classification and Characteristics of Army Stress Casualties During Operation Desert Storm," *Hospital and Community Psychiatry*, Vol. 43, No. 8 (1992), pp. 812–815.

[25] Solomon, Noy, and Bar-On, "Risk Factors."

prominent problem, given the limited levels of teamwork required for many support duties and the fact that reserve forces spend comparatively little time training with comrades. In addition, both types of units most likely have limited confidence in their combat skills and weapons training in comparison to regular infantry units. Support units, especially in brief conflicts, that transition to combat operations may already be physically challenged with the fatigue and sleep loss inherent in the deployment of forces. Meanwhile, reserve forces are likely to face more home-front worries such as concerns about family welfare and lost income. Evidence from the Gulf War suggests that long deployments help units of all varieties overcome many of these deficits.[26]

In addition, elite units generally report low levels of psychiatric casualties. For example, in World War II the 442nd Regimental Combat Team was the most decorated unit in the U.S. Army. During the Italian campaign, this unit had almost no psychiatric casualties.[27] In addition, psychiatric casualties as a percentage of total casualties rarely exceed 5 percent for the 81st, 101st, and 17th Airborne Divisions committed to the European theater.[28] Low stress casualty rates were also reported for elite units in the Yom Kippur and Lebanon wars and the Falklands War.[29] Factors that may protect these units include increased cohesion due to high retention of members, intense training and the self-respect that comes from being viewed as elite, as well as briefer combat engagement periods than conventional divisions.

---

[26] K. M. Wright et al., *Operation Desert Shield Desert Storm: A Summary Report* (Washington, D.C.: Walter Reed Army Institute of Research, 1995).

[27] Also relevant, however, may be that the unit, made up of Japanese Americans, enjoyed cultural homogeneity.

[28] A. L. Hessin, "Neuropsychiatry in Airborne Divisions," in W. S. Mullins and A. J. Glass (eds.), *Medical Department, United States Army Neuropsychiatry in World War II*, Vol. 2, *Overseas Theaters* (Washington, D.C.: U.S. Government Printing Office, 1973), pp. 375–398.

[29] F. D. Jones, "Psychiatric Lessons of War."

Battlefield Contributors to Stress

> Why is it that before the storming of a fort, or fighting a battle, men are thoughtful, heavy, restless, weighed down with care? Why do men on these occasions ask more fervently than usual for the divine guidance and protection in the approaching conflict. . . . For all my poor comprehension may tell, tomorrow I may be summoned before my maker.
>
> —*John Shipp, British soldier and three-time volunteer for the forlorn hope at Bhurtpore, India, 1826*[30]

Anticipation

Fear of the unknown can be a significant source of stress for combat soldiers. In the Vietnam War, individuals scheduled for combat were observed to report an increase in vague medical complaints with no identifiable physical causes. In addition, a psychiatrist stationed with the 25th Infantry Division noted that when deployment to Vietnam was uncertain, there was an increase in psychiatric referrals and complaints that decreased once deployment became definite.[31] The WRAIR study of Gulf War stressors noted that among ground force soldiers after the beginning of the air war, "each perceived delay in movement forward lowered morale, each anticipation of 'starting the job' raised it." Moreover, "morale re-ascended to an extremely high level as soon as there was firm knowledge that the Ground War was to begin." Soldiers reported in interviews that the start of the ground war was the "greatest stress reliever of the months of deployment."[32]

---

[30] J. Shipp, *The Path of Glory* (London, 1969, p. 64), referenced in Holmes, *Redcoat*, p. 385.

[31] A. W. Johnson, "Combat Psychiatry, Part 2: The U.S. Army in Vietnam," *Medical Bulletin of the U.S. Army Europe*, Vol. 25 (1969), pp. 335–339.

[32] Wright et al., *Operation Desert Shield Desert Storm*, p. 46.

Combat Intensity

Combat intensity is the greatest battlefield predictor of stress reactions. In general, as the number of physical casualties rise, so will the number of CSRs. This same relationship exists with disease and nonbattle injuries (DNBI), a classification that includes psychiatric casualties (Figure 3.1). In addition, the rates of combat stress reactions have been correlated with combat intensity independent of physical casualties. Following the Lebanon War, after-action reports of several engagements were rated and ranked for combat intensity and other factors, blind of casualty estimates. Factors used by the panel included preparation, type of battle, adequacy of support, enemy resistance, and commander's relation to higher command. As seen on Table 3.2, the ranking of combat intensity predicted both physical casualty rates and CSR rates.

Figure 3.1
Daily Casualty (Wounded and Killed in Action) and DNBI Rates Among U.S. Marine Infantry Units During the Okinawa Operation

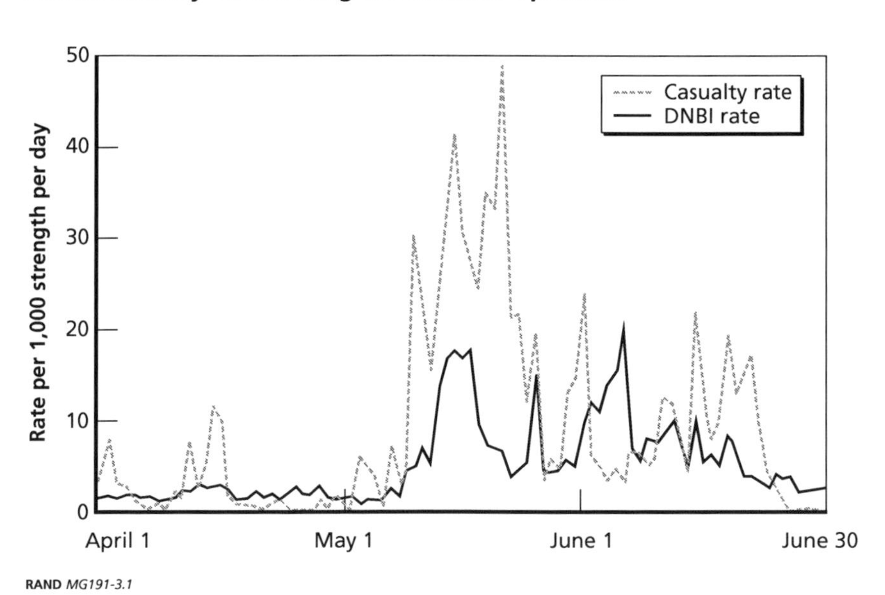

RAND *MG191-3.1*

SOURCE: Adapted with permission from C. G. Blood and E. D. Gauker, "The Relationship Between Battle Intensity and Disease Rates Among Marine Corps Infantry Units," p. 342.

Table 3.2
Combat Intensity Predicts Combat Stress Reactions in the Lebanon War, 1982

Ranking	Number of Physical Casualties (KIA + WIA)	Number of Psychiatric Casualties	Ratio
1	36	31	86:100
2	32	9	39:100
3	10	1	10:100
4	12	0	00:100

KIA = killed in action. WIA = wounded in action.

NOTE: Battalions were ranked from most (1) to least (4) stressful on five factors:

- Preparation (enemy location, mission, false alarms, training).
- Type of battle (artillery, air attack, ambush, hostage, mine field).
- Support (tactical, logistical, materiel).
- Enemy resistance (strong, adequate, weak).
- Trust by commander in higher command (justified pressure, some pressure, adequate support).

SOURCE: Adapted from G. L. Belenky, C. F. Tyner, and F. J. Sodetz, *Israeli Battle Shock Casualties: 1973 and 1982*, p. 33.

Different levels of combat intensity can produce different types of stress casualties. For example, reanalyzed World War II data showed that massed exposure to intense combat produced more psychiatric casualties in those who remained out of combat for only a short time. In contrast, sporadic stress, as evidenced by few combat days per calendar days, resulted in more disciplinary casualties who remained out of combat for significant periods of time.[33] This factor was most likely in play in the latter years of the Vietnam War, where forces were rarely engaged in combat and yet 60 percent of all medical evacuations were neuropsychiatric, especially drug and alcohol abuse.

[33] Noy, "Battle Intensity and the Length of Stay on the Battlefield as Determinants of the Type of Evacuation."

Duration

Infantry soldiers are most at risk for stress reactions during their initial forays into battle and following prolonged periods of combat duty. From World War II, it has been estimated that green troops are at their highest risk of breakdown during the first 5 to 21 days of combat and that veteran troops risk collapse sometime between 30 and 250 days of combat. The most frequently cited estimate of a soldier's emotional lifespan, calculated by Beebe and Appel, is between 80 and 90 days of combat.[34] Alternatively, the British Army claimed its soldiers lasted considerably longer, largely because of a practice of withdrawing men from the line for rest.[35] This is in contrast to U.S. forces, which committed men to the line almost perpetually.

Type of Battle

> [Man] lay flat in the trench, shaking and groaning and moaning that he was afraid to die and half shrieking as each shell came by.
>
> —*Lieutenant in the 4th Worcesters at the Somme*[36]

Anecdotal data from a number of 20th-century conflicts suggest that conditions on a moving battlefield, either in attack or retreat, are more hospitable to a soldier's sanity than are those on a static battlefield. A psychiatrist with the 4th Armored Division in September 1944 noted that a gasoline shortage halted aggressive action, allowing the Germans to gain the initiative and counterattack.

---

[34] Beebe and Appel estimated that half of combat infantrymen studied in the Mediterranean theater of operations broke down after 85–90 days of company casualty days of combat. Beebe and Appel, *Variation in Psychological Tolerance*.

[35] R. H. Ahrenfeldt, *Psychiatry in the British Army in the Second World War* (London: Routledge & Kegan Paul Ltd., 1958).

[36] William Strang, *Diaries* (Liddle Collection, University of Leeds), referenced in Shephard, "Shell-Shock on the Somme," p. 51.

> On 16 September the incidence of exhaustion casualties began to increase and by 27 September had reached an alarming figure. On that date 71 men in the division became exhaustion casualties, including a high percentage of non-commissioned officers . . . Between 15 September and 30 September there were 355 incidences of exhaustion—more than had occurred in the previous two months.[37]

Similarly, the psychiatrist Peter Bourne noted that "Neuropsychiatric casualties in combat occur predominantly when the lines of battle are static and diminish sharply when the troops are on the move, even though they may be in full retreat."[38]

Several factors make defensive or "passive" operations particularly stressful to troops. First, troops engaged in these types of operations often become unwitting receptacles for sustained artillery or air attacks. This was certainly the case in World War I, where, for example, an Australian soldier at the Battle of Pozieres counted 75 shells landing within five minutes on a four-acre parcel of land. He recalled, "All day long the ground rocked and swayed backwards and forwards from the concussion . . . . Men were driven stark staring mad and more than one of them rushed out of the trench over towards the Germans. Any amount of them could be seen crying and sobbing like children their nerves completely gone."[39] Artillery and air attacks were also the cause of a number of stress casualties in the Lebanon War.[40] Another factor is the sense of helplessness engendered in troops waiting for the enemy to choose to attack. Helplessness is also a major factor when receiving indirect fire and possibly biological or chemical attacks. At best, the typical soldier can search for shelter or put on protective gear. Paradoxically, active engagement with the en-

---

[37] Marlowe, *Cohesion, Anticipated Breakdown, and Endurance*, p. 5, quoting E. W. Mericle.

[38] P. G. Bourne, "Military Psychiatry and the Vietnam Experience," *American Journal of Psychiatry,* Vol. 127, No. 4 (1970), p. 482.

[39] P. Charlton, *Pozieres* (London: Leo Cooper, 1986). C.E.W. Bean, *Official History of Australia in the War of 1914–1918,* Vol. 3, *The AIF in France 1916* (Canberra: Australian War Memorial, 1929), referenced in Shephard, "Shell-Shock on the Somme," p. 54.

[40] Noy et al., "Mental Health Care."

emy, with its cathartic release of anger and therapeutic sense of distraction and control, may be less stressful.

Physical Hardships

Physical hardships include hunger, thirst, sleep deprivation, and extremes in weather. The fact that food, water, sleep, and regulation of body temperature are the tenets of forward psychiatric restoration suggests that these factors play a pivotal role in the development of combat exhaustion.[41] Also, a number of anecdotal reports from World War II and the Korean War suggest that harsh environmental conditions such as rain and frigid temperatures, along with sleep deprivation and a lack of nutritious food, precipitated stress reactions.[42]

> When the average soldier goes into combat he is usually rested, well fed, and able to withstand the normal emotional stresses of combat. As time goes on he becomes increasingly tired and less nourished, and with this decrease in physical well-being there is a corresponding decrease in his ability to cope with emotional stress . . . . Thus, physical fatigue operates by lowering the soldier's ability to withstand emotional stress.[43]

In addition, the deleterious consequences of sleep loss are well known. Sleep deprivation is known to impair cognitive capacities and levels of sustained vigilance.[44]

---

[41] Belenky interview.

[42] E. W. Mericle, "The Psychiatric and the Tactical Situations in an Armored Division," *United States Army Medical Department Bulletin,* Vol. 6, No. 3 (1946), pp. 325–334. E. C. Ritchie, "Psychiatry in the Korean War."

[43] F. Hanson, "The Factor of Fatigue in the Neuroses of Combat," *United States Army Medical Bulletin,* Vol. 9 (1949), pp. 147–150.

[44] D. R. Haslam and P. Abraham, "Sleep Loss and Military Performance," in G. L. Belenky (ed.), *Contemporary Studies in Combat Psychiatry* (Westport, CT: Greenwood Press, 1987), pp. 167–184.

Summary

In summary, an array of factors contribute to the development of combat stress reactions. Personality has not been shown to contribute to battle fatigue. The presence of home-front stressors appears to be strongly related to breakdown. Level of education is also a factor, but the extent to which this applies to a highly educated military is unclear.

Factors related to morale and unit assignment can also influence the appearance of stress reactions. Units that are highly cohesive, and whose members have a high degree of confidence in their military prowess and trust their commander, have lower rates of stress reactions than units that lack such benefits. Infantry soldiers are at considerable risk of developing stress casualties, but combat service and support and reserve units that incur casualties will have disproportionately high rates of psychiatric cases. Members of elite units seem well protected, at least in the short term, from the psychological consequences of war.

Battlefield contributors to stress are numerous. The anticipation of battle appears to be very anxiety producing. Once the battle starts, soldiers are at their greatest risk of developing stress reactions during their initial forays into battle and following long-term and unremitting exposure. Combat intensity is also a critical factor. Stress casualties should be expected during and after engagements that result in significant numbers of wounded and killed in action (WIA and KIA). In addition, battles that are characterized by static warfare and/or indirect fire likely instill a sense of helplessness on the combatants. Finally, physical hardships such as extremes in weather or physiological deprivations such as extreme hunger, thirst, or sleep deprivation will also increase the risk of battle fatigue.

CHAPTER FOUR

Stress in the City: An Evaluation of the Risk of Combat Stress Reactions in Urban Warfare

> Now I knew without a doubt that we were in deep shit. For a lingering, uncomfortable moment, I felt as though we had accidentally stumbled into the twilight zone. The view was radically different from any other that I had experienced in Vietnam. We had walked through an invisible curtain from an achingly green, vividly living world, into a black and white madness of destruction and death.
>
> —*Nicholas Warr* [1]

On its face, the urban environment would seem to be a highly stressful combat venue. In a three-dimensional battle space, enemy forces can position themselves in subterranean structures, at ground level behind buildings and doorways, and in buildings on virtually any floor (including rooftops). Taking into account the myriad of structures present in most urban environments, there is an incalculable number of lines of sight and fields of fire. Friendly forces, which typically must ultimately advance on ground level, can be engaged from virtually any conceivable direction, even from the one whence they came.

The urban battle is often an up-close and personal one. Many targets are within a 50-meter range. Often, advancing forces must clear individual buildings that can contain multiple floors of enemy

[1] N. Warr, *Phase Line Green: The Battle for Hue, 1968* (New York: Ivy Books, 1997), p. 54.

combatants. Friendly forces that enter each room may face the barrel of an enemy's rifle, a cowering civilian, or a comrade in arms. These soldiers will have split seconds to decide whether or not to engage. Enemy forces masked in civilian clothes will complicate this decision even further.

The structural composition of urban environments presents a number of other obstacles to friendly forces. Ubiquitous rubble creates significant obstacles to advancing troops and increases the risk of injuries. Tanks often have a limited capacity to engage basement and roof-level targets given limits on the elevation and depression of their weapons. Moreover, buildings will often block the trajectories of tank shells, artillery, and ordnance dropped by fixed-wing aircraft. Even when such ordnance is on target, the subsequent explosion can have secondary effects on civilians and infrastructure.

In addition to the factors described above, many other features are present that would appear to engender high rates of stress casualties. For one, urban operations are expected to inflict a high casualty toll, thus increasing rates of stress casualties that are known to correspond to rates of WIA. In addition, support personnel, already acknowledged to be at increased risk for stress casualties, may find themselves in closer proximity to combat operations than is elsewhere the norm.[2] Individual infantry or support units may also be dispersed across an entire block or throughout a single building. Such dispersal can infringe upon unit cohesion and limit the amount of support available to any given soldier.[3] Given that leadership would be similarly dispersed, units that have trained under the dominant leadership of a single company commander, for example, must now rely more heavily on those at lower leadership echelons such as platoon leaders, squad-level sergeants, or fire-team leaders. An individual's faith in leadership becomes tested to the extreme. Critically, these factors will operate in conjunction with any weaknesses in training and morale.

---

[2] R. W. Glenn, S. L. Hartman, and S. Gerwehr, *Urban Combat Service Support Operations: The Shoulders of Atlas* (Santa Monica, CA: RAND Corporation, MR-1717-A, 2003).

[3] LTC Carl Castro, Ph.D., interview with the author, Washington, D.C., May 20, 2003.

For a variety of reasons, then, troops engaged in urban operations would seem to be at considerable risk for combat stress reactions.

Although one might surmise that stress reactions would be greater in urban operations, at present it is unclear whether this is indeed the case, due to the absence of prior multi-battle investigations. Such an investigation would be important for a variety of reasons. First, it is necessary in estimating the combat troop strength required to sustain long-running urban operations. Casualty rates are typically high in this environment, and unforeseen depletion in combatant manpower can pose a significant danger to remaining forces and mission success. Similarly, estimation of stress casualties is necessary so that both combat and medical support units will be able to anticipate and consequently plan for the type and severity of stress reactions. In the following analysis, we attempted to evaluate the subjective stress inherent in urban operations and the corresponding risk of stress casualties.

A View from the Ground

Because they were there, combat-trained soldiers and marines provide a unique insight into the stressors inherent in urban operational settings. We interviewed a number of soldiers and marines with urban operational experience about the stress inherent in urban operations. Their responses attest to the inordinate degree of stressors present in the urban environment.

For example, one veteran from the Battle of Hue in Vietnam stated that he has often been asked to describe fighting during that war. He notes that his description centers on Hue.

> [Before Hue] there were long periods of abject boredom interrupted abruptly by moments of sheer terror and violence. And the long period could be days, weeks, months of day after day patrols and ambushes and nothing happens in terms of combat and then when you least expect it shit happens . . . . When I de-

> scribe combat in Hue City, nobody was ever bored; it was non-stop, all-out war at very close range all day long.[4]

The comments of other Hue veterans share a remarkable similarity. Factors central to the stressors they experienced included their lack of training in urban warfare, the "constant combat," the terrain that confined their movements and restricted their visibility, the "myriad places for [enemy] cover/concealment," and the close proximity of combatants. Accordingly, LtGen (ret.) Christmas asserts that

> The restricted environment, the noise or din of intense urban fighting, the debris effect which causes many if not most of your casualties, and the closeness of your opponent, usually within 35 meters, which often leads to hand to hand fighting all make for stressful days and nights.[5]

When directly queried about the comparative stress of urban operations versus operations on other types of terrain, most veterans from Hue readily concluded that urban-based operations were notably more stressful.

Panama combat veteran COL (ret.) Johnny Brooks found that urban fighting during Operation Just Cause was particularly stressful:

> There are so many dangers in urban areas that normally do not exist in open terrain: non-combatants, close proximity of enemy, hidden routes of egress and ingress, potential for sabotage, et al. The real problem is that around every corner a concealed difficulty (enemy or otherwise) may exist. There is little to no respite. The unknowns just put significant stress on all soldiers. On open terrain in most cases you can see a lot more area than on urban terrain. There is just a sense of uncertainty in urban areas that does not exist in open terrain. On open terrain, normally you can find some place where you can put eyes in all directions

---

[4] Nicholas Warr, interview with the author, Solona Beach, California, May 9, 2003.

[5] LtGen (ret.) George Christmas, written comment to the author, June 6, 2003.

> and feel somewhat secure that you are safe. In urban areas, you can't see past the next wall. [It is] definitely more stressful.[6]

U.S. marines serving in Operation Restore Hope in Somalia recall an urban-based stability operation where the line between civilian and foe was blurred. LtCol (ret.) John Allison, although wounded in the Gulf War, readily noted that actions in Mogadishu, and peacekeeping actions in Los Angeles following the 1992 riots, were significantly more stressful than the 1991 Gulf War.[7] LtCol (ret.) Robert Barrow, who served in Beirut and Mogadishu, agrees:

> It's three-dimensional. People can be below in sewers, above in buildings and in alleys and doorways and can blend in and out of the environment easily. You can go into urban areas and there can be bad guys and good guys and a lance corporal has to determine if he can pull the trigger and there are a lot of difficult decisions: Do I shoot or not shoot? In that regard it's difficult. And the terrain: it's been bombed and blown up and you can clear it (of enemy forces) . . . and it doesn't mean next time you go back it's cleared again. . . . It's not like World War II where cities were evacuated. Every time you had to go back to that area and look at it from a threat situation, and every time I drove back to those areas . . . it was mentally exhausting just from scanning and being on the look out because you're vulnerable. It's difficult. People blend in. There are no distinct uniforms.[8]

The Impact on Combat Stress Reaction

Urban operations no doubt induce a significant degree of stress in their participants, but the question remains as to whether such stress translates into actual increased incidence of acute stress reactions. While most medical professionals argued that there were no data on

---

[6] COL (ret.) Johnny W. Brooks, written comment to the author, May 6, 2003.

[7] LtCol (ret.) John Allison, interview with the author, Arlington, Virginia, June 5, 2003.

[8] LtCol (ret.) Robert Barrow, interview with the author, Tampa, Florida, June 16, 2003.

which to base their opinions, many of these same professionals were willing to hypothesize that urban combat would indeed result in an increased CSR rate.[9] Beyond personal opinion, the perceived risk of stress casualties in urban operations is noted more formally in other sources. For example, Army field manual entitled *Leaders' Manual for Combat Stress Control* states:

> Combat in built-up areas will be unavoidable in war and operations other than war (conflict). Units will have to plan for attack and defense in urban areas and for fluid battles around them. The usual static, house to house nature of urban warfare, with many snipers, mines, and booby traps, tends to increase battle fatigue casualties unless troops are well-trained and led.[10]

In addition, Gregory Ashworth, in his book *War and the City*, writes that

> the type of battle dictated by the urban environment imposes particularly severe strains on those subjected to it. The continuous high level of alertness demanded by close actions, the physical discomfort, and the insecurity of isolated small unit operations without fixed lines, secure flanks, or protected rear-all contribute to the rapid onset of battle fatigue within hours rather than days.[11]

Consequently, just as those who have had their "boots on the ground" argue that the subjective experience of stress is greater in urban operations, it seems also that many other authorities believe that such stressors may translate into higher numbers of CSR.

---

[9] David H. Marlowe, Ph.D., interview with the author, Alexandria, Virginia, May 21, 2003. LTC Elspeth C. Ritchie, M.D., interview with the author, Washington, D.C., June 3, 2003. CDR Jack S. Pierce, M.D., interview with the author, Arlington, Virginia, May 20, 2003. COL James Stokes, M.D., interview with the author, Fort Sam Houston, Texas, March 19, 2003.

[10] Department of Defense, *Leaders' Manual for Combat Stress Control*.

[11] G. J. Ashworth, *War and the City* (New York: Routledge, 1991), p. 121.

A Look at the Numbers

In an effort to objectively determine the risk of stress casualties in urban warfare, we evaluated the rates of acute stress reactions in prior urban conflicts. What follows is a review of available data from the battles of Brest, Manila, Hue, Panama, and the Task Force Ranger (TFR) October 3, 1993 engagement in Mogadishu. Data are also provided from the urban battles of Aachen, Seoul, Lebanon, and Chechnya, though with a caveat about their validity.

World War II

The Battle for Brest

The urban portion of the Battle of Brest was waged from September 8 to September 18, 1944. In this battle, 50,000 U.S. troops fought 30,000 German defenders of Brest and its outlying countryside. The German authorities had evacuated most of the city's original 80,000 French inhabitants. Ordered to defend the city to the last man, German fighters, despite preparatory aerial and artillery bombardment of the city, staged a stiff resistance.[12] Street fighting was described as intense, and Germans "seemed to contest every street, every building and every square" and "machine gun and antitank fire from well-concealed positions made advances along the thoroughfares suicidal."[13]

The Ninth U.S. Army Medical Section's after action report details the available evidence of neuropsychiatric casualties in the form of monthly casualty figures and hand-drawn graphs.[14] Important to the interpretation of these data are the specific dates in which two separate American divisions fought within the city. To this end, the

[12] M. Blumenson, *The European Theater of Operations, Breakout and Pursuit* (Washington, D.C.: Center of Military History, United States Army, 1989).

[13] Ibid., p. 646.

[14] *After Action Report, Medical Section, Ninth U.S. Army, Period 5–30 September 1944 Inclusive* (National Archives and Records Administration, Record Group 407).

2nd Infantry Division's (ID) urban combat experience was most intense from September 8 to 14 with the subsequent four-day period consisting primarily of a tactical pause followed by a final attack on September 18.[15] The overall month's psychiatric casualty rate for the 2nd ID was 95 with 530 WIA, resulting in 17.9 psychiatric cases per 100 WIA. The Ninth Army's figure (Figure 4.1) for stress casualties suggests that rates remained relatively constant throughout the month with no dramatic increase during the battle's most intense phase (September 8–14).

In contrast, the 8th ID fought within the city limits for only a three-day period beginning on September 8, 1944. On September 10 this division was transferred from Brest to the nearby Crozon Peninsula, where it engaged German forces on open terrain from September 15 to 19. This change allows a comparison (albeit imperfect) between operations on urbanized and open terrain. Over the entire month of September this division suffered approximately 16.4 stress casualties for every 100 WIA. Figure 4.2, depicting the Ninth Army's figure, seems to illustrate no dramatic increase in neuropsychiatric casualties for the three days of city fighting, and casualty rates for these days do not appear to be more pronounced than those engendered during the Crozon portion of the operation. For September 8, 9, and 10, stress casualty rates appear to approximate 8, 20, and 8 percent respectively. In contrast, during operations on the Crozon Peninsula, stress casualty rates approximate 13, 14, 22, and 8 percent for September 15–18 respectively.

While estimating raw numbers from the Ninth Army's graph involves some approximating, casualty figures for both the 2nd and 8th IDs do not indicate that stress casualties were particularly high during the urbanized portion of the fight for Brest.

---

[15] C. Lawrence and R. Anderson, *Measuring the Effects of Combat in Cities: Phase 1* (Annandale, VA: The Dupuy Institute, 2002). We would like to thank the authors of this report for their helpful comments.

Figure 4.1
Daily Admissions to the Clearing Station for the 2nd Infantry Division

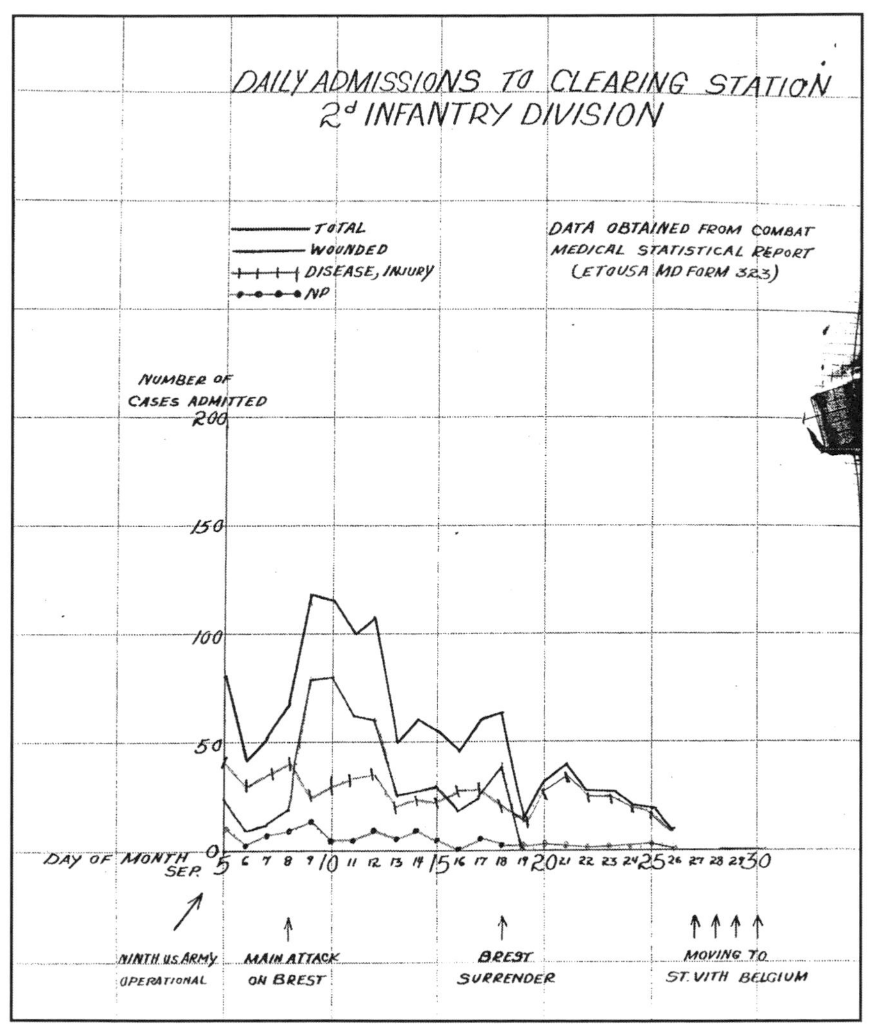

RAND *MG191-4.1*

SOURCE: After Action Report, Medical Section, Ninth U.S. Army, Period 5–30 September 1944 Inclusive, National Archives and Records Administration, Record Group 407.

NOTE: The relatively high rates of "Disease and Injury" casualties found in these figures may result from cold and rainy weather of the time.

Figure 4.2
Daily Admissions to the Clearing Station for the 8th Infantry Division

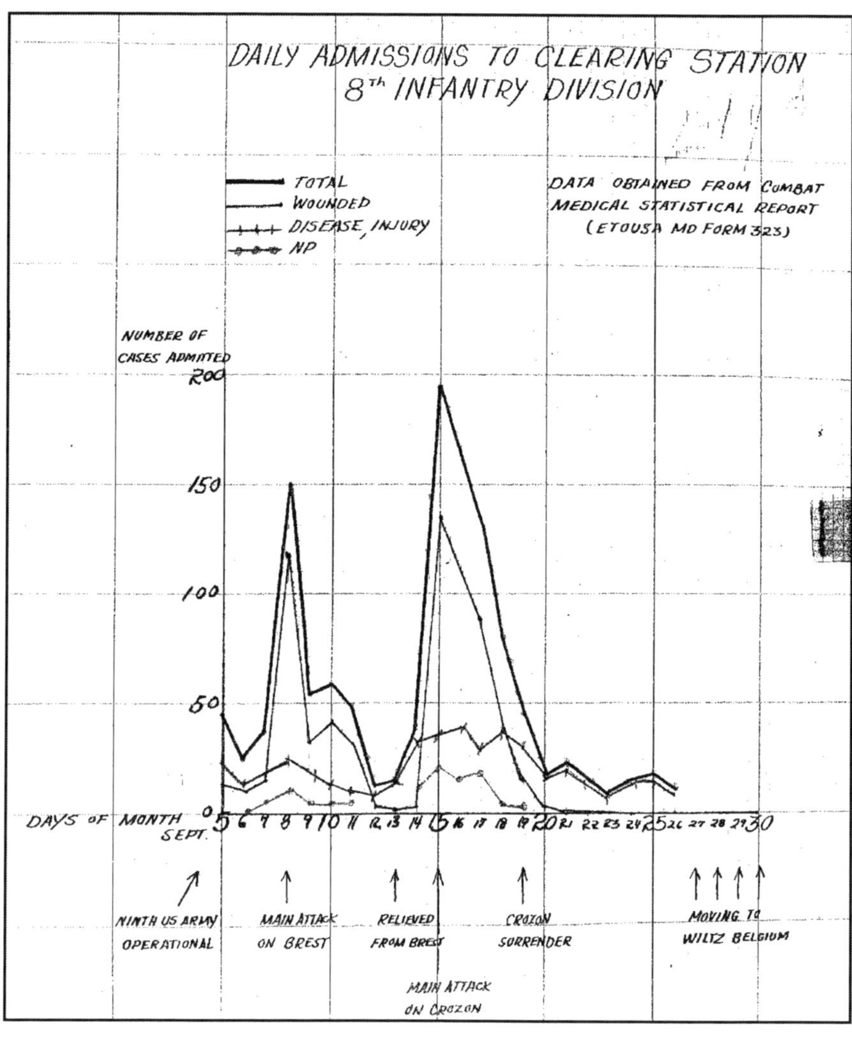

RAND *MG191-4.2*

SOURCE: After Action Report, Medical Section, Ninth U.S. Army, Period 5–30 September 1944 Inclusive, National Archives and Records Administration, Record Group 407.

Battle of Manila

The Battle of Manila was fought from February 3 to March 3, 1945. The city, with an area of 14.5 square miles, contained a prewar population of 1 million. Although the Japanese army was willing to cede the city to U.S. forces, the Imperial Navy ordered its defense. The Japanese force was small (17,000) and was deficient in artillery, armor, air support, and communications equipment. It was also ill trained and lacked doctrine for operating on restricted terrain. The Japanese nonetheless waged a strong defense, utilizing dismounted naval machine guns and cannon on multiple presighted fields of interlocking fire. In addition, mine fields restricted the maneuverability of U.S. armor. Unconventional tactics, such as defending civilian-populated buildings and opening fire on U.S. forces while pretending to surrender, were also utilized.[16]

The U.S. assault with the 37th ID and 1st Cavalry Division was initially constrained by rules of engagement that prevented aerial bombardment and strafing runs. Artillery bombardments were also restricted during the battle's first week.[17] Once artillery fires were permitted, building assaults were prepared with tank and artillery fire, with tanks subsequently guiding dismounted infantry to the building's entrance. Once inside, U.S. soldiers fought room to room using light suppressive fires, grenades, flame-throwers, and bazookas to clear out enemy forces.[18] American casualties for this operation were 1,010 KIA and 5,565 wounded.[19]

---

[16] R. R. Smith, *The War in the Pacific: Triumph in the Philippines* (Washington, D.C.: Center of Military History, United States Army, 1991).

[17] Prior to the lifting of artillery restrictions, the 37th ID suffered an average of 26 KIA per day. After they were lifted, the KIA rate dropped to an average of six per day.

[18] T. M. Huber, *The Battle of Manila* (Fort Leavenworth, KS: Combat Studies Institute, Command and General Staff College, 2002), http://www.globalsecurity.org/military/library/report/2002/MOUTHuber.htm (accessed September 22, 2003).

[19] Smith, *The War in the Pacific.*

Data on stress casualties come from a historical report by the 37th ID's clearing company, 112th Medical Battalion,[20] and from the division psychiatrist's report.[21] All data apply to the 37th ID alone. Before detailing the casualty figures, it is relevant to note that the division appeared to undergo a change in the way it handled stress casualties just prior to the Luzon campaign. Central to this change was an attempt to keep all mild to moderate stress casualties at regimental aid stations where they would be held for a short time and then returned to some form of duty. In contrast, more severe casualties were sent to the clearing hospital, where a psychiatrist would diagnose combat exhaustion.[22] As the division psychiatrist notes, "This fact will introduce a statistical fallacy to the over-all report on admission percentages submitted by the Division Psychiatrist" because only the group sent to the clearing station became recorded as battle fatigue casualties.

With this in mind, between January 9 and March 4, 1945, a time generally encompassing the actions in Manila,[23] the 37th ID recorded only 64 stress casualties. With an approximate total of 3,057 casualties (KIA and WIA), the ratio of stress to physical casualty was a misleadingly low 2.1:100. Given the use of regimental aid stations as a holding facility for psychiatric casualties, it is impossible to determine the actual rate of neuropsychiatric casualties. Because it is unclear whether similar casualty sorting procedures were in place during earlier campaigns, it is difficult to make comparisons to the division's previous stress casualty rates during the Bougainville campaign (12 percent) and New Georgia campaign (25 percent).

---

[20] *Headquarters, 29th Infantry Division* (National Archives and Records Administration, Record Group 407).

[21] *Report of the Division Psychiatrist, 37th ID, Luzon Campaign* (National Archives and Records Administration, Record Group 407).

[22] The 129th Infantry regimental aid station, for example, treated a total of 140 soldiers, with "some degree of psychiatric inadequacy." Of this total, 20 soldiers (14.3 percent) were admitted to the 112th's clearing station.

[23] Casualties prior to the Battle of Manila (January 9–February 4, 1945) were not comparably heavy—111 KIA and 439 WIA—and thus most of the 64 stress casualties can be presumed to originate in Manila's city limits.

Although raw data are not presented for the three regimental aid stations, it is likely that the division psychiatrist would have been aware of the numbers in a general way, and it is also likely that he would have been told if the numbers of incoming stress casualties were exorbitant. No concern over high stress casualties is mentioned at any point in his report. In fact, he gives several reasons why the numbers were low. These reasons include "excellent leadership and training of the men" and "increased hatred of the enemy," as many soldiers appeared anxious to "avenge the killing of their buddies or the torturing of the Filipino civilians which they themselves had witnessed." Morale was also high.

> Most soldiers stated that for the first time since they have been overseas, they have seen for what they are fighting; liberation of the prisoners of war and civilian interests and the Filipino peoples' appreciation and gratitude for liberation.[24]

Finally, one factor not observed by the psychiatrist but noteworthy nonetheless was a rotation policy instituted by commanders where units with 14 days of heavy fighting at the line were rotated out of the line of contact.[25] Leader actions such as this can play a pivotal role in reducing the morbidity associated with combat stress.

The Battle of Hue

The Battle of Hue took place from January 30 to March 2, 1968. During the 1968 Tet offensive, Vietcong and North Vietnamese regulars infiltrated the lightly defended city with a force of 20 battalions. The marines dispatched to evict the North Vietnamese forces fought across three fronts, within the densely populated walls of the Citadel, in the Gia Hoi district to the Citadel's east, and in the southern portion of the city where the government office building, hospital, and city schools were located. The operation consisted of a month-long, house-to-house battle whose success was measured in

---

[24] *Report of the Division Psychiatrist, 37th ID, Luzon Campaign.*

[25] Huber, *The Battle of Manila*.

yards per day. This difficult battle was made more challenging by the marines' lack of urban operational training and the initially restrictive ROE, intended to save the historic city's infrastructure, that prevented the use of artillery and supporting air cover. In the marines' favor was a civilian population who were either evacuated to local refugee camps or remained cloistered in their basements. These civilians generally did not engage or interfere with U.S. military operations. In addition, the North Vietnamese army never successfully infiltrated beyond the line of contact or the frontal assault waged by the marines. As a result, the marines' rear area remained safe.[26] Finally, many marines noted that fighting largely stopped during the night, allowing combatants on both sides to sleep.

Estimates of stress reactions stem from a report by Blood and Anderson[27] detailing disease and non-battle injuries for Hue, in which DNBI rates were tallied for the period of Operation Hue City and the month before and after the battle.[28] DNBI data are also presented for separate marine battalions from the period of May 1 to August 31, 1968, and for the assault on Okinawa during World War II (April 1 to June 31, 1945).

Results of this analysis demonstrate that the DNBI rate was 0.92/1,000 strength/day for the month before Hue, 0.98/1,000 strength/day during Hue, and 0.92/1,000 strength/day in the month following Hue (Figure 4.3).

As seen, these numbers are much lower than those found for the 90-day Battle of Okinawa, which resulted in a DNBI rate of 4.56/1,000 strength/day (see Figure 4.4). When evaluated as a ratio to WIA, the rates were 5.6:100 and 69.4:100 for the battles of Hue and

---

[26] J. H. Willbanks, *The Battle for Hue, 1968* (Fort Leavenworth, KS: Combat Studies Institute, Command and General Staff College, 2002), http://www.globalsecurity.org/military/library/report/2002/MOUTWilbanks.htm (accessed September 20, 2003).

[27] C. G. Blood and M. E. Anderson, "The Battle for Hue: Casualty and Disease Rates During Urban Warfare," *Military Medicine,* Vol. 159, No. 9 (1994), pp. 590–595.

[28] In a separate communiqué, the author, Christopher Blood, noted that hospital ship records suggested that Hue's DNBI rate by and large reflects actual rates of battle fatigue.

Figure 4.3
DNBI and Total Casualty Rates for the Period of and Surrounding the Battle for Hue

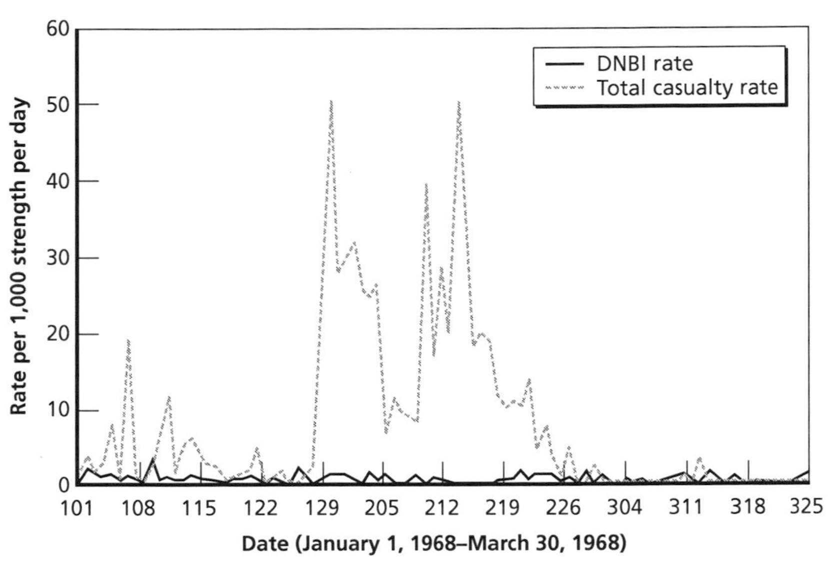

RAND *MG191-4.3*

SOURCE: Adapted with permission from C. G. Blood and M. E. Anderson, "The Battle for Hue: Casualty and Disease Rates During Urban Warfare," p. 594.

Okinawa respectively. In addition, the DNBI ratio for Hue was even lower when compared with those among infantry battalions during a different four-month period of the Vietnam War that averaged a remarkably high 71.2:100 WIA (Figure 4.5). Finally, it is noteworthy that in contrast to the Battle of Okinawa, DNBI rates in the Battle of Hue do not vary as a function of casualty rates.

Blood and Anderson's findings, reviewed above, did not illustrate the previously demonstrated relationship between battle intensity and DNBI rates. The authors noted that prior research on DNBI rates suggests that the tempo of the Vietnam War may have weakened the relationship between DNBI and combat intensity. Such a pattern mirrors that found throughout the Vietnam War, where battle fatigue rates never correlated with casualty rates.

Figure 4.4
DNBI and Total Casualty Rates for Marines During the Battle of Okinawa, 1945

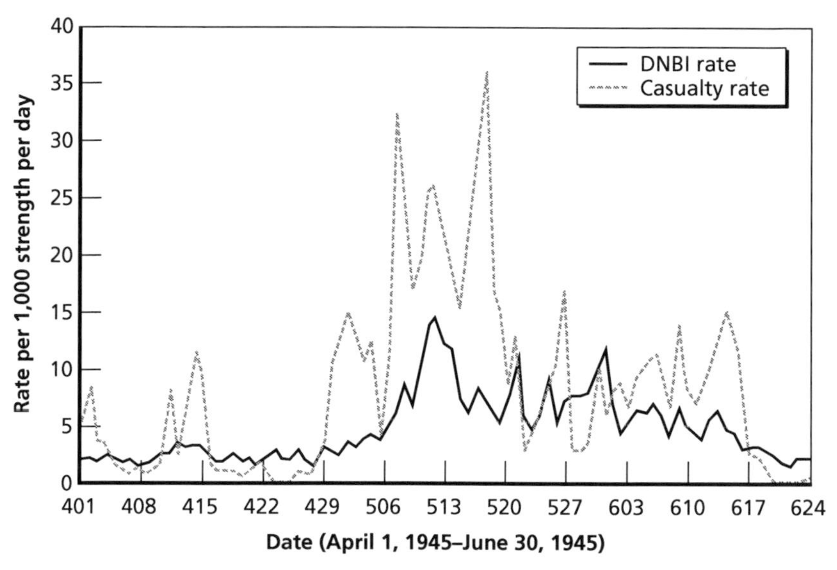

RAND *MG191-4.4*

SOURCE: Reprinted with permission from C. G. Blood and M. E. Anderson, "The Battle for Hue: Casualty and Disease Rates During Urban Warfare," p. 594.

Another potential hypothesis may be that the prolonged and intense combat experienced by marines in Hue gave them a greater sense of purpose than they had previously experienced out in the field where engagements were often initiated by enemy forces. Hue commanders echoed such sentiments. For example, retired Brigadier General Downs noted that "Morale in our unit was good, I always attributed it to engaging the enemy and winning and experiences up to that were getting sniped at from afar, booby trapped . . . and [you] feel you're not accomplishing anything. This time we saw the enemy, we were with them."[29]

---

[29] BGen (ret.) Michael Downs, interview with the author, Quantico, Virginia, February 12, 2003.

Figure 4.5
DNBI and Total Casualty Rates for Separate Marine Infantry Battalions During an Alternative Four-Month Period of the Vietnam War

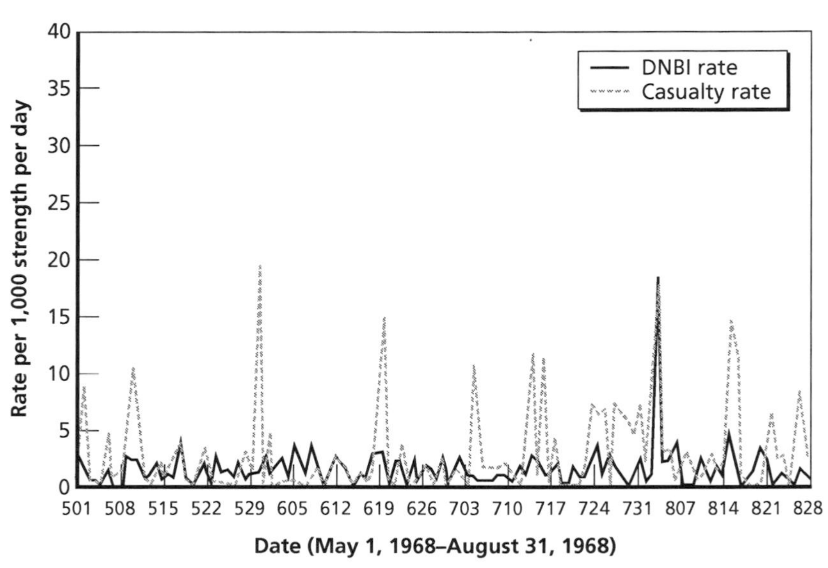

RAND *MG191-4.5*

SOURCE: Reprinted with permission from C. G. Blood and M. E. Anderson, "The Battle for Hue: Casualty and Disease Rates During Urban Warfare," p. 594.

More Limited-Duration Urban Conflicts

Panama

On December 20, 1989, U.S. forces invaded Panama in Operation Just Cause with the goal of forcibly removing dictator Manuel Noriega. Approximately 27,000 U.S. troops, consisting of a mix of conventional and special operations forces, attacked 27 Panamanian targets in Panama City and Colon. These forces fought 3,500 Panamanian Defense Force (PDF) infantry troops. While fighting occurred in such areas as Paitilla Airfield, the International Terminal at Torrijos Airport, and Ancon Hill, the brunt of combat operations took place at the Panamanian Commendancia. U.S. casualties consisted of 23 KIA and 324 WIA, while PDF casualties have been estimated at approximately 450.[30] Collateral damage was generally limited.

A key source of information on stress casualties for this operation comes from an interview with Dr. James Stokes (COL, USA MC).[31] According to Stokes, the rates of acute CSR for this operation were not particularly high. In fact, Dr. Stokes knows of only one case, which originated from the battle for the Commendacia. In this case a medic was evacuated to his company area, was given useful duties, and was subsequently functioning well. Dr. Stokes assumes that most overstressed soldiers were managed similarly. In addition, there was only one known soldier with classic combat exhaustion, who was brought with the wounded to the only Air Force mobile staging facility at the airfield. He was given reassurance and an opportunity to sleep, after which he asked to return to duty.[32]

Several factors most likely contributed to the low rate of formal battle fatigue casualties. First, U.S. forces had anticipated and planned for this operation for a period of two years with extensive urban operations training. Critically, the overall fight lasted only a day, with specific firefights rarely lasting more than several hours. PDF forces suffered from poor morale and leadership. Most did not fight, choosing instead to leave their posts.[33] Thus, many relevant

---

[30] L. A. Yates, *Operation Just Cause in Panama City, December 1989* (Fort Leavenworth, KS: Combat Studies Institute, Command and General Staff College, 2002), http://www.globalsecurity.org/military/library/report/2002/MOUTYates.htm (accessed September 20, 2003).

[31] Stokes interview.

[32] In a written communication, Dr. Stokes recounted another important observation from Operation Just Cause. Soldiers from Ranger and Airborne units experienced "minor but temporarily disabling orthopedic injuries" from one of their difficult night jumps. Evacuated to San Antonio, Texas with the seriously wounded, many felt ashamed that they had "let down" their comrades. Instead of outright evacuation, many soldiers would have probably volunteered to "man stationary defensive positions around the airfield." After the fighting they could return to their units or pursue medical treatment with their pride and honor intact. Dr. Stokes observed, "This case was used successfully to defend the deployment of austere medical holding facilities at brigade level and in the Future Force, which can also be used for CSR casualties."

[33] Yates, *Operation Just Cause*.

battlefield stressors were not present. Moreover, many U.S. troops were special operations units who by the nature of their extensive training and high unit cohesion were afforded a significant degree of protection from CSR.

Task Force Ranger

In the summer of 1993, the Joint Special Operations Task Force, or "Task Force Ranger," was deployed to Mogadishu, Somalia. Its mission was to capture or destroy the command infrastructure of Mohammed Aidid's militia that had ambushed and killed 23 Pakistani troops. TFR conducted six raid missions beginning in August 1993, but its largest and best-known mission occurred on October 3 of that year. TFR seized the Olympic Hotel and apprehended the targeted leadership faction. In the process, however, two Black Hawk helicopters were shot down over militia-controlled territory. A planned 30-minute combat operation turned into a 14-hour fight for survival. In the end, 19 American soldiers were killed, 77 were wounded, and one was held as a POW.

There are no published accounts regarding the extent of acute combat stress reactions during or following this operation. Some indication is available from interviews. Most notably, COL Larry Lewis, a command psychologist with TFR, observed that there were "very, very few" battle fatigue casualties.[34] The only formal stress casualty known to these authors occurred following the operation, when a hospital-based medic succumbed to the stress of treating incoming casualties. After a day's rest, the medic returned to duty.[35] In accounting for the "very, very few" stress casualties, COL Lewis echoed the above-referenced observation that special operations soldiers such as Rangers, given their highly cohesive nature and high training tempo, do not easily succumb to psychiatric breakdown.[36]

---

[34] COL Larry K. Lewis, Ph.D., interview with the author, Fort Meade, Maryland, August 21, 2003.

[35] COL John Holcomb, M.D., interview with the author, Houston, Texas, March 4, 2003.

[36] While the 10th Mountain Division also participated in this operation through staging the battle zone evacuation of TFR, rates of stress casualties for this division are not known.

For the reasons cited above, it appears unlikely that Operation Just Cause and the TFR engagement on October 3, 1993, serve as models of what might be expected if a large conventional force faced a prolonged and life-threatening battle in urbanized terrain. They have value with respect to shorter actions involving urban terrain and those involving elite forces.

Battle of Jenin, Israel

The Battle of Jenin was waged from the early morning of April 3 to April 10, 2002. In response to the Palestinian second intifada, Israeli Defense Forces (IDF) launched raids into a number of Palestinian-controlled areas including the refugee town of Jenin, known to some as "the suicide-bombing capital of the West Bank." The Israelis utilized a mix of special forces and reserve and active duty conventional forces that entered the city in tanks and armored personal carriers. They faced stiff resistance from Palestinian gunmen who began preparing the camp's defenses as early as the Passover massacre in Netanya on March 22, 2002. Specific defensive tactics employed by the Palestinians included the use of snipers, booby-trapped doorways, suicide bombers, and ambushes. Palestinian civilians were also present in high numbers, and enemy fighters did not distinguish themselves with uniforms. In one particular ambush, only three Israeli reservists from a 16-man unit remained unharmed after they were targeted by a bomb and swept with small arms fire.[37] All told, there were just under 100 Israeli casualties consisting of 23 KIA and 60 WIA. Interestingly, despite the battle's brevity, there were a total of 17 combat stress casualties. This resulted in a ratio of CSR to total casualties of 20.5:100. Alternatively, when looked at as a function of WIA, the ratio becomes 28.3:100.[38]

---

[37] J. Hammer, "A War's Human Toll: Israel Wins a Fierce Battle, but the Victory Gives Birth to Another Saga of Blood and Fire," *Newsweek,* April 22, 2003.

[38] BG Gideon Avidor (IDF, ret.) presented to members of the 10th Mountain Division command and staff, Fort Drum, NY, January 13, 2003. This presentation provided statistics indicating 23 percent killed, 60 percent WIA, and 17 percent mental. News reports indicate 23 Israeli KIA. It is consequently assumed that the numbers of WIA and CSR were 60 and 17, respectively.

A Secondary Analysis

Four other battles were reviewed for this report, but because of insufficient or unclear data, they were considered best relegated to a separate analysis. These battles include Aachen, Seoul, the Lebanon War, and Chechnya.

The Battle of Aachen took place between October 10 and 21, 1944. Daily rates of combat exhaustion indicate that during the period of city fighting the ratio of stress casualties to WIA was 30:100, a value considerably higher than the typical 20 percent maintained throughout most of the division's World War II history.[39] Only two battalions out of three infantry regiments actually participated in the fight. Given the availability of only division-wide medical statistics, it was impossible to ascertain the CSR contribution of the two battalions that fought in the city. In addition, an infantry regiment fighting outside the city reported its "toughest fighting" of the war.[40] This regiment may have contributed significantly to the overall exhaustion rates, further skewing the division-wide data.

With regard to the Battle of Seoul, Albert Glass has written, "the 1st Marine Division, which bore the brunt of the fighting for Seoul, suffered heavy battle losses and consequently incurred a large number of psychiatric casualties."[41] However, the validity of this statement is questionable for several reasons. First, although the Seoul campaign lasted from September 20 to October 5, 1950, the actual period of city fighting lasted a total of only four days (September 25–28) and

---

[39] *Headquarters, First Medical Battalion, APO 1, United States Army* (National Archives and Records Administration, Record Group 407).

[40] *History of the 18th Infantry (1st ID) for the Period 1–31 October 1944* (National Archives and Records Administration, Record Group 407).

[41] A. J. Glass, "History and Organization of a Theater Psychiatric Services Before and After June 30, 1951," in *Recent Advances in Medicine and Surgery (19–30 April 1954): Based on Professional Medical Experiences in Japan and Korea 1950–1953*, Vol. II (Washington, D.C.: Army Medical Service Graduate School, Walter Reed Army Medical Center, Medical Science Publication No. 4, 1954), p. 361.

was waged by a single infantry regiment.[42] The exact period of time and the exact constitution of troops Dr. Glass was referring to are unknown. In addition, the only psychiatric-related mention in the division surgeon's after action report is that "ineffectives were few."[43] The term "ineffectives" seems to pertain to psychiatric casualties.

The Israelis invaded Lebanon in 1982. The war was fought in both mountainous and urban terrain. Evidence of stress casualty rates for the Lebanon War stem from a report written by COL Yheskel Besser.[44] COL Besser tabulated physical and psychiatric casualties according to whether they occurred on urban or nonurban terrain. His report observes that CSR casualties constituted 4.5 percent of physical casualties for urban terrain and 7.4 percent for nonurban terrain.[45] Unfortunately, the accuracy of these numbers is unclear, given that they are far below the 23 percent CSR rate known for the overall war.[46] For this reason, both the surgeon general[47] and the chief psychologist[48] for the IDF during the Lebanon campaign suggest that these numbers are likely an unreliable indicator of the true psychiatric casualty rate.

The Chechen war lasted from December 1994 to August 1996. The war was predominantly fought in the cities, most notably in the city of Grozny. Unfortunately, there are no known published reports on the overall psychiatric casualty rate for Chechnya or the Battle of

---

[42] J. H. Alexander (ed.), *Battle of the Barricades: U.S. Marines in the Recapture of Seoul* (Washington, D.C.: Marine Corps Historical Center, 2000).

[43] Annex Queen to 1st Marine Division Special Action Report, Division Surgeon, August 28 to October 7 (Marine Corps History and Museum Division, Korean War, CD #1).

[44] COL Y. Besser, "Military Operations in Urbanized Terrain—Medical Aspects—Lebanon War 1982: A Case Study" (unpublished paper, August 1985).

[45] The numbers of individuals with a physical injury without CSR, a physical injury with CSR, and CSR only were 471, 17, and 5 for urban terrain and 638, 25, and 24 for nonurban terrain, respectively.

[46] Belenky, Tyner, and Sodetz, *Israeli Battle Shock Casualties*.

[47] BG Eran Dolev, M.D. (IDF, ret.), interview with the author, London, England, March 20, 2003.

[48] COL Ron Levy, Psy.D. (IDF, ret.), interview with the author, Ksarsaba, Israel, April 7, 2003.

Grozny in particular. The only evidence of the psychological consequences of combat stem from a report by Major General V. S. Norvikov, of the Russian medical service, who during the war surveyed 1,312 soldiers with respect to psychiatric symptoms.[49] According to a translated summary of his findings, 72 percent of surveyed soldiers reported psychological problems such as "insomnia, lack of motivation, high anxiety, neuro-emotional stress, tiredness, and hypochondriacal fixation."[50] In addition, problems such as "asthenic depression, a weak, apathetic or retarded motor state" were reported by 46 percent of the sample, and 26 percent exhibited "psychotic" reactions with associated features such as "high anxiety or aggressiveness, a deterioration of moral values or interpersonal relations, excitement or depression."[51] In addition, following the war, a number of veterans were apparently plagued with PTSD, or what was termed the "Chechen Syndrome."

Unfortunately, these data are extremely difficult to interpret. While the report indicates that 1,312 troops were screened, it is not clear how these troops were chosen and whether their experiences were representative of most infantry forces in Chechnya. Also, it is never stated how many soldiers were deemed combat ineffective or were evacuated from the battle lines. It may be that the levels of insomnia or anxiety reported were not sufficiently disturbing so as to impair combat performance or necessitate evacuation. The 26 percent psychotic rate, if referring to individuals experiencing auditory or visual hallucinations or delusions, also seems curiously high. The author does state that "the percentage of troops with combat stress disorders was higher than in Afghanistan," but absent the context of a defini-

---

[49] V. S. Norvikov, "Psycho-Physiological Support of Combat Activities of Military Personnel," *Military Medical Journal* (Russia), No. 4 (1996), pp. 37–40, referenced in T. L. Thomas and C. P. O'Hara, "Combat Stress in Chechnya: The Equal Opportunity Disorder" (Fort Leavenworth, KS: Foreign Military Studies Office), http://fmso.leavenworth.army.mil/fmsopubs/issues/stress.htm (accessed December 2, 2004).

[50] Thomas and O'Hara, "Combat Stress in Chechnya," p. 5. Hypochondriacal fixation, according to the paper, refers to a soldier who becomes fixated on cardiovascular function, worrying, for example, about heart attacks or breathing problems.

[51] Thomas and O'Hara, "Combat Stress in Chechnya," p. 5.

tion of "stress disorder" and its relation to total casualty rates, this too remains uninterpretable.

Summary

Based on data from the Battles of Brest, Manila, and Hue, it appears that fighting within city limits does not necessarily result in increased rates of stress casualties. The Battle of Jenin was the only battle in which rates of stress reactions seemed relatively high given its brief duration. These data contrast sharply with both the hypothesized CSR rates and what might be expected given the comments of veterans of urban combat interviewed in support of this study. Why do these discrepancies exist? Several hypotheses are reviewed below.

Limitations of the Review

First, it must be considered that limitations of the historical literature review introduced a degree of bias. The main conclusions rely on the first three battles reviewed, Brest, Manila, and Hue. Of these three, Hue was the only one where the findings were objectively determined. That is, unit records were surveyed for daily reports of DNBI casualties. In contrast, no such records were available to the authors for Brest and Manila. Conclusions of stress casualty rates from Brest relied on hand-drawn graphs created by the Ninth Army's medical section. Such a depiction is subject to human error and bias, though probably not to any significant degree. Similarly, conclusions for Manila are based solely on the division psychiatrist's account of the battle's casualties. Division psychiatrists are probably motivated to put a "best face" on reports that may bear some indication of their job performance. In addition, it is noteworthy that because of the scarcity of combat stress data, only a very limited number of recent urban battles could be reviewed for this report. Thus, it is possible that many other battles exacted a greater psychiatric toll than the ones reviewed for this analysis. The general lack of high stress casualties found in this study, however, does lend credence to our conclusions.

Hypothesized Protective Factors for Urban Combat Operations

One hypothesis that must be considered is that combatants did experience a high degree of stress in these operations but that stress did not translate into increased stress reactions. This might be the case, if other factors serve to protect the mind from breakdown. For example, the 37th ID's psychiatrist noted that soldiers at the Battle of Manila experienced an increase in morale due to the liberation of civilians and American POWs. In addition to operational idiosyncrasies, some mitigating factors may be shared across the entire spectrum of urban operations.

Infantrymen, due to their close contact with enemy soldiers, may experience an increased sense of tactical control during urban operations. As previously noted, some commanders from Hue observed an increase in morale due to the newfound ability to see and engage enemy combatants. As previously cited, BGen (ret.) Michael Downs noted that "Morale in our unit was good . . . . This time we saw the enemy, we were with them."[52] Similarly, Col Meadows (ret.) stated that "close combat and close awareness of enemy was for us a bit of a morale booster."[53]

Urban combat operations are inherently close-quarters fights. Despite the presence of multidirectional threats, such proximity allows the average rifleman to inflict damage on enemy soldiers he would otherwise be unable to achieve and also to see the effects of his actions. This certainly precludes soldier feelings of helplessness and provides a sense of control over their environment. This is in marked contrast to fighting on terrain in which artillery and booby traps often play a much a greater combat role. Under such circumstances, riflemen lack the ability to personally respond effectively to the attacks or casualties inflicted on fellow unit members. Similarly, aside from Israeli forces in Beirut, enemy artillery fire has not played a significant role in the post–World War II urban battles considered here.

---

[52] Downs interview.

[53] Col (ret.) Chuck Meadows, interview with the author, Bainbridge, Washington, February 18, 2003.

In contrast, small arms carried much greater weight during these urban operations. In this last regard, the psychiatrist Albert Glass writes:

> Very obviously, if you raise the destructive power of the weapon so that the individual cannot cope with it, then non-effectiveness is enhanced. If you have a weapon that is of minor destructive power such as bows and arrows, or rifles, more people can cope with it. This is why men tell you in combat they don't mind small arms fire; what they detest is artillery fire or mortar or other high explosives. So if you diminish the destructiveness, your curve looks different; if you raise it, then you have more non-effective people.[54]

Another factor relevant to all of the battles reviewed in this chapter is that stress casualties were evaluated with respect to members of the offensive forces. Reviewing the relative risk of offensive versus defensive forces, Reuven Gal and Franklin Jones write:

> in defensive operations, especially with impending danger but without active engagement to break the tension, the soldier is subjected to an enforced passivity and experiences a feeling of helplessness. By contrast, in offensive operations, even though the risk may be greater, the soldier is active, has a vicarious sense of control over the situation, and is distracted from personal concerns.[55]

Finally, urban combat operations with their multidirectional threats may require constant vigilance and consume significant quantities of mental energy. According to Michael Downs, "my mind was engaged from the time I opened my eyes in the morning to time they closed at night. The danger [was] all around. You get more tired from

---

[54] A. J. Glass, "Leadership Problems of Future Battle: Presented to The U.S. Army War College" (Carlisle Barracks, PA, 1959), referenced in F. D. Jones, "Traditional Warfare Combat Stress Casualties," p. 41.

[55] R. Gal and F. D. Jones, "A Psychological Model of Combat Stress," in F. D. Jones, L. R. Sparacino, V. L. Wilcox, J. M. Rothberg, and J. W. Stokes (eds.), *War Psychiatry* (Washington, D.C.: TMM Publications, 1995), pp. 133–48, 141.

that then you do from physical activity."[56] Such vigilance should limit the opportunity for inner-focused fears and anxiety. A British soldier describing his World War I experiences wrote:

> But I did not feel afraid, or at least not nearly so afraid as I had felt immediately before going over . . . But now there was so much to think about, so much to distract my attention, that I forgot to feel afraid—it is the only explanation. The noise, the smoke, the smell of gunpowder, the rat-tat of rifle and machine-gun fire combined to numb the senses. I was aware of myself and others going forward but of little else.[57]

Application Toward Future Urban Conflicts May Be Limited

Many recent urban operations involve populated city landscapes that tend to involve a number of irregular characteristics less prominent in those World War II and Vietnam War events. Primary among these characteristics is the trend for enemy forces to choose to fight in an environment replete with civilian inhabitants. The day in which the defending force willfully evacuates or does not deliberately put the civilian population at risk may be past. This creates a host of problems for friendly forces. One significant problem is the limitation such a population poses to U.S. aerial and artillery firepower. The desire to minimize civilian loss of life and collateral damage to the extent now practiced is a fairly recent phenomenon. In addition, enemy combatants who forswear uniforms easily blend into civilian environments. This makes engagement decisions extremely difficult and increases the risk of killing innocents. Enemy forces can easily disguise themselves as noncombatants, thus increasing the probability that they can penetrate the line of contact and operate in the friendly force's rear area. Finally, depending on the host city's cultural mandates, the presence of civilian populations increases the risk of friendly forces receiving direct fire from demographic groups typically

---

[56] Downs interview.

[57] J. Ellis, *Eye-Deep in Hell* (London: Croom Helm, 1976), referenced in Kellett, *Combat Motivation*, p. 288.

considered noncombatants (e.g., women, children, and the elderly). Directly targeting these populations, as well as accidentally engaging noncombatants, may pose unique psychological risks to professionally trained soldiers and marines.

This last point is noteworthy. During our interviews, one of the factors most frequently suggested to cause psychological problems was the battlefield presence of civilian noncombatants. Dr. Neil Greenberg observed, "in cities [soldiers] encounter more noncombatants than combatants. Also some places in [the] world you see child soldiers, what do you do about that? [It's the] conflict of having to kill a child. I think the civilian aspect is all important."[58] Similarly Dr. Jack Pierce stated that "you're being shot at by children, teenagers, men dressed as women; you're also shooting at them. That's a higher stress."[59] According to Dr. Elspeth Ritchie, having to wound or kill civilians would likely place an individual at "risk for both acute and chronic symptoms."[60] While we are aware of no single example of CSR secondary to such an incident, there are several examples of civilian shootings appearing to engender post-traumatic stress disorder. For example, at least two known cases of PTSD that followed Operation Just Cause were related to an incident in which soldiers mistakenly killed a car full of civilians at a roadside checkpoint.[61] Similar problems occurred when an ambulance driver in Somalia was ambushed and, after the subsequent firefight, believed he had killed two Somali children.[62]

The battles reviewed in this report where the operating environment involved a large civilian presence include the Battle of Manila, Operation Just Cause, the battle waged by TFR, and the Battle of Jenin. Unfortunately, the reliance on special forces during TFR limits its general application to regular force troops. Likewise, the

---

[58] LtCdr Neil Greenberg, M.D., interview with the author, London, England, May 5, 2003.

[59] Pierce interview.

[60] Ritchie interview.

[61] Stokes interview.

[62] Ritchie interview.

brief duration of Operation Just Cause limits its general application. However, such conventional forces were utilized in Manila and Jenin. While civilians in Manila caused the creation of restrictive ROE and were used as shields by Japanese servicemen, they did not pose any direct threat to friendly forces and they did not obscure actions of uniformed Japanese soldiers. In contrast, Jenin exhibits nearly all of the negative consequences of civilian-populated battlegrounds. Stress casualties for this battle reached 28.3 for every 100 WIA. Unfortunately, the timing of this report did not permit an evaluation of the recently fought Operation Iraqi Freedom (OIF). The phases of major combat and peacekeeping operations both involved an enemy who utilized tactics similar to those described above. Given these factors, our report cannot equivocally address the risk of stress casualties during unconventional urban operations.

CHAPTER FIVE

Reversing the Tide: Treatment Principles for Battle Fatigue

The stress-related risks of future military operations, including those on urban terrain, require that members of the U.S. military learn the knowledge and skills relevant to the treatment and prevention of stress casualties. Acute combat stress reactions can pose a significant burden to a tactical unit in terms of lost manpower. However, a number of steps are available to both combat and medical support units that can reverse the symptoms of battle fatigue and limit the permanent loss of afflicted soldiers and marines. It is especially important that commanders and NCOs familiarize themselves with these steps, as their initial actions may have the most influence on the outcome of a given combat stress casualty. The following review describes ways to identify and treat battle-fatigued soldiers with a special focus on within-unit restoration techniques. The success of PIES is also reviewed. The term battle fatigue will be liberally used in this chapter, as this is the battlefield relevant name for CSR.

Symptoms and Diagnosis

> One of the most characteristic traits of neurotic reactions to battle is the manner in which the symptoms alter with the lapse of time, change of geographic setting, distance from the combat scene and progress or lack of treatment. What begins as a severe anxiety reaction in

the combat area may end up as a severe depression in a rear area or at home.

—*Grinker and Spiegel* [1]

There are many challenges inherent in the identification and diagnosis of combat stress reactions. Frequently, multiple symptoms are present in a single casualty (polymorphic), and any two casualties may manifest completely different types of symptoms. Symptoms also change in their nature and severity over time (lability). These factors make CSR identification and diagnosis difficult even for trained psychiatrists. The process becomes even more challenging when it is members of the combat unit who must make the diagnostic decision. With the exception of the medic, line personnel lack clinical experience and must simultaneously focus on operational realities. The functional definition of CSR relating to combat ineffectiveness is also relative. It is dependent on many factors, including the various attitudes toward stress reactions held by the unit or commanding officers, the extent to which the unit believes it can spare a given soldier, and the capability of the unit to care for the casualty.[2]

Table 5.1 lists a number of common battle fatigue symptoms. A study of 100 Israeli stress casualties from the Lebanon War helps discern the extent to which these types of symptoms are seen on the battlefield and highlights some common symptom constellations.[3] For example, 41 percent of the CSR patients showed only one symptom, the most common of which were disabling forms of anxiety (13 percent of the entire sample) and depression (9 percent). Other problems included dissociation (lost awareness of self; 6 percent), somatic com-

---

[1] Grinker and Spiegel, *Men Under Stress*, p. 83.

[2] Z. Solomon, N. Laor, and A. C. McFarlane, "Acute Posttraumatic Reactions in Soldiers and Civilians," in B. A. van der Kolk, A. C. McFarlane, and L. Weisaeth (eds.), *Traumatic Stress: The Effects of Overwhelming Experience on Mind, Body, and Society* (New York, London: The Guilford Press, 1996).

[3] T. Yitzhaki, Z. Solomon, and M. Kotler, "The Clinical Picture of Acute Combat Stress Reaction Among Israeli Soldiers in the 1982 Lebanon War," *Military Medicine*, Vol. 156, No. 4 (1991), pp. 193–197.

plaints (physical complaints such as headaches with no known cause; 5 percent), disturbances in states of awareness (4 percent), psychosis (reports of auditory or visual hallucinations; 3 percent), and changes in motor activity (1 percent). In addition, 48 percent of CSR patients had two or more symptoms (a "polymorphic" clinical picture), e.g., 8 patients demonstrated both anxiety and depression symptoms, while 5 patients had both anxiety and somatic complaints. Finally, the clinical picture in 11 percent of casualties was labile, that is, symptoms changed over the course of time. In nearly half these cases the soldier presented with anxiety only to end in a state of depression.

Table 5.1
Mild and Severe Signs and Symptoms of Combat Stress Reaction

	Mild	Severe
Physical symptoms	Trembling Jumpiness Cold sweats, dry mouth Insomnia Pounding heart Dizziness Nausea, vomiting, or diarrhea Fatigue "Thousand yard" stare Difficulty thinking, speaking, and communicating	Constantly moves around Flinches or ducks at sudden sound and movement Shakes, trembles Paralysis Inability to see, hear, or feel Physical exhaustion Freezes or is immobile under fire Staggers or sways Panics, runs under fire
Emotional symptoms	Anxiety, indecisiveness Irritability, complaining Forgetfulness, inability to concentrate Nightmares Easily startled by noise, movement, and light Tears, crying Anger, loss of confidence in self and unit	Talks rapidly and/or inappropriately Argumentative, acts recklessly Indifferent to danger Memory loss Stutters severely, mumbles, or cannot speak Insomnia, severe nightmares Sees or hears things that do not exist Rapid emotional shifts Socially withdrawn Apathetic Hysterical outbursts Frantic or strange behavior

SOURCE: Adapted from Army FM 6-22.5, *Combat Stress*.

Memorization of the various symptoms may be neither necessary nor helpful given that the fog of war will most certainly thwart a thoughtful battlefield diagnosis. What is important is that fellow soldiers and leaders are able to identify behavioral changes suggestive of deterioration in soldiers or marines. The U.S. Army's Field Manual 6-22.5, *Combat Stress,* notes that

> Any service member who shows persistent, progressive behavior that deviates from his baseline behavior may be demonstrating the early warning signs and symptoms of a combat stress reaction. Trying to memorize every possible sign and symptom is less useful to prompt diagnosis than to keep one simple rule in mind: Know your troops, and be alert for any sudden, persistent or progressive change in their behavior that threatens the functioning and safety of your unit.[4]

Treatment

Doctrinal Echelons of Treatment

Table 5.2 and Figure 5.1 present the sublabels of battle fatigue and the doctrinally based echelons at which CSR or battle-fatigued soldiers can be treated. Various sublabels of battle fatigue simply indicate the echelon at which the afflicted soldier can be treated. These echelons are dependent on both the soldier's symptoms and the tactical situation of the soldier's unit. "Light" battle fatigue applies to those soldiers whose apparent symptoms are mild and where the unit is capable of caring for them. Individuals with severe symptoms may also qualify for "light" battle fatigue, provided the symptoms respond quickly to helping actions. In contrast, "heavy" battle fatigue cases are those that physicians or mental health staff refer for further evaluation. This may happen if the soldier's condition is too disruptive to the unit's mission or if there is reason to have the soldier evaluated for

[4] Department of Defense, *Combat Stress* (Department of the Army, FM 6-22.5, June 23, 2000). Army FM 6-22.5 is a joint U.S. Army, Navy (Navy Tactics, Techniques, and Procedures [NTTP] 1-15M), and Marine Corps (Marine Corps Reference Manual [MCRP] 6-11C) manual.

Table 5.2
Description of Various Sublabels for Battle Fatigue

Battle Fatigue Sublabel	Symptoms and Treatment Locale
Light	Treatment within the unit. Symptoms are mild, or severe symptoms respond quickly to helping actions.
Heavy	Treatment in a medical facility. Casualty evacuated from unit because symptoms disruptive to unit mission.
Duty	Seen by physician and immediately cleared for duty.
Rest	Sent to unit's nonmedical CSS unit for rest.
Hold	Held for treatment in triager's own medical facility until symptoms permit RTD.
Refer	Referred and transported to more secure medical facility.

SOURCE: Adapted from Army FM 22-51, *Leaders' Manual for Combat Stress Control.*

another medical condition. If a physician decides to hold the soldier for further treatment within the medical unit, then it is at this time that the battle fatigue "case" becomes a casualty. The terms "duty," "rest," "hold," and "refer" indicate the options available to the attending physician (see Table 5.2). In general, soldiers and marines should be treated at the least restrictive level and in the greatest proximity to the original unit as both the tactical situation and the symptom severity warrant. The nature of the treatment provided by medical authorities and the unit are detailed below.[5]

Treatment by Mental Health Units

When an individual with combat fatigue is evacuated out of his or her unit, the basic principles underlying treatment are described by the terms proximity, immediacy, expectancy, and simplicity, or PIES. Servicemen are ideally treated behind—but as close to—the line of contact as is feasible given the tactical situation and condition of the individual. Treatment should be provided as soon after evacuation as

---

[5] Ibid.

Figure 5.1
Diagram of Sorting Choices and Labels for Battle Fatigue Cases

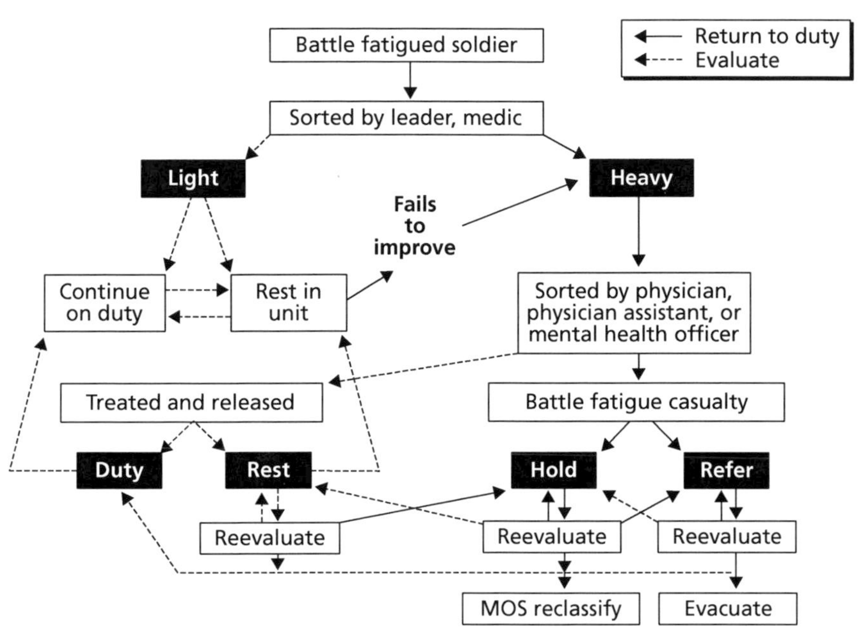

RAND *MG191-5.1*

SOURCE: Adapted from Army FM 22-51, *Leaders' Manual for Combat Stress Control.*

possible, and casualties are told by an authoritative source that they are expected to recover and return to their units. The overall treatment approach is simple. Soldiers are provided replenishment in the form of warm food and drink, and they are allowed time to rest and sleep in a relatively safe environment. They are treated separately from physical casualties so as not to communicate that they are ill, and ideally sedatives and tranquilizing medications are avoided. To the extent possible, unit members and leaders are encouraged to visit the soldier and tell him he is needed and welcomed back to the unit. In some cases, casualties undergo a reconditioning program, where they participate in physical and military training geared toward rebuilding their confidence and identity as a soldier. In both cases, they are encouraged to talk about the battlefield experiences that precipitated their breakdown. In addition, prior to the provision of treatment, battle fatigue cases are medically screened to rule out other medical and psychiatric conditions that require other treatment.[6]

An alternative means of describing battle fatigue treatment is through the four "R"s: *Reassurance* of a quick recovery from a confident and authoritative source; *Respite* from intense stressors; *Replenishment* in the form of water, a hot meal, sleep, regulation of body temperature, and hygiene; and *Restoration* of perspective and confidence through conversation and work.[7]

Various treatment echelons are available to medical authorities so that casualties who are unresponsive to the initial restoration attempt or are deemed to have a poor prognosis can be evacuated further up the rearward zone. The goal of treatment is to return the soldier to his original unit. The tactical situation permitting, this can be done at any point in the treatment process.[8] The success of this approach for a relatively stable battle space is reviewed below.

The Role of Expectation

> The more [soldiers] are treated like hospital patients, the worse they get.
>
> —*Ingraham and Manning*[9]

PIES, in addition to replenishing the soldier physically, also seeks to harness the stress casualty's desire to return to his unit and to reassert his identity as a soldier. What keeps a soldier at the line, braving both death and dismemberment, are his comrades. The desire to not let his buddies down is a powerful one. When this desire is nurtured and sustained it is a compelling force in motivating a soldier with battle

---

[6] F. D. Jones, "Traditional Warfare Combat Stress Casualties."

[7] Paraphrased from Stokes, written comments.

[8] F. D. Jones, "Traditional Warfare Combat Stress Casualties."

[9] L. H. Ingraham and F. J. Manning, "Psychiatric Battle Casualties: The Missing Column in a War Without Replacements," *Military Review*, August 1980, pp. 18–29, 24.

fatigue to either remain in the unit or leave the safe confines of the support hospital to return to that unit.

Both the proximity of treatment to the battlefield and the immediacy with which it is begun following evacuation influence these opposing forces. The psychiatrist Franklin Jones writes that "expectancy is the central principle from which the others derive. A soldier who is treated near his unit in space (proximity) and shortly after leaving it (immediacy) can expect to return to it. Distance in space or time decreases this expectancy."[10] Fast-moving lines of contact or battles waged across deep operating environments often serve to increase the evacuation distance, thus compromising the principles of proximity and immediacy.

Treatment simplicity also relates to a soldier's expectation to return to duty. Complex treatments such as the World War I–era narcosynthesis or electroshock "may only strengthen the soldier's rationalization that he is ill physically or mentally."[11] Expressing pity or overconcern for the individual may also engrain his pathology. Such was the case where a Survival, Evasion, Resistance, Escape school student began to experience a conversion reaction in the form of severe back pain and lockjaw. As medical authorities swarmed about this presumed patient, the commander matter of factly took the soldier out for a walk and insisted they talk about his experience. The student quickly recovered from his debilitating back pain and was able to continue the program.[12]

Limitations in the Application of Current CSC Doctrine

> Once you go to medical, you're out of your unit and the chances of returning to your unit, because they move so fast, is very unlikely.
>
> —*CDR J.S. Pierce, M.D.*[13]

---

[10] F. D. Jones, "Traditional Warfare Combat Stress Casualties," p. 9.

[11] F. D. Jones, "Psychiatric Lessons of War," p. 10.

[12] Morgan interview.

[13] Pierce interview.

> You don't just want a talking head, you want someone with face validity whom they can identify with and who has seen the elephant and someone who knows what the elephant looks like. I think one of the mistakes we made is not to bring in the people with the kind of line experience and credentials who can make the message real to folk who really don't want to believe it.
>
> —*David H. Marlowe, Ph.D.*[14]

There are two significant limitations on the application of current combat stress control doctrine that merit attention. The first problem stems from recent-day reliance on maneuver warfare. In the Yom Kippur War,[15] the Gulf War, and Operation Iraqi Freedom, forward units moved so fast that the distances between forward elements and medical treatment facilities were large.[16] This not only makes it impossible for line personnel to visit battle-fatigued members of their unit, but also makes it extremely difficult to return recovered soldiers back to duty. Despite this problem, mental health units have partly adapted to maneuver warfare by deploying preventive teams with maneuver brigades where PIES interventions can be facilitated at the unit level.[17]

Second, most combat stress control units or other mental health assets are not organic to the tactical or support units they serve.[18] As a result, members of mental health units are unknown and oftentimes

---

[14] Marlowe interview.

[15] Levy interview.

[16] Marlowe interview.

[17] In Operation Iraqi Freedom's major combat phase, both division mental health services and combat stress control units advanced into Baghdad with the lead U.S. Army maneuver brigades. These units provided PIES and debriefing interventions to small units in the active combat area, without having to hold stress casualties for treatment. Stokes, written comments.

[18] This is a fault not of CSC doctrine but of U.S. Army policy, priorities, and manpower constraints. Stokes, written comments.

not trusted by the line community.[19] According to a recent report, soldiers experiencing significant distress during OIF were three times more likely to turn to a fellow soldier in their own unit for help than to formal mental health assets or chaplains.[20]

While some suggest that CSC units are sufficiently flexible to adapt to the changing realities of warfare,[21] others have observed that the current configuration of mental health support is inadequate.[22] In particular, one medical authority has suggested that forward psychiatry based on the PIES approach requires "radical reconfiguration to match current operational realities."[23] This same individual asserts that the reconfiguration will have to adapt to a military less reliant on divisions and instead "deploy forward psychiatric resources as organic to maneuver units" at the brigade level. Importantly, these necessary changes are already taking shape. Based on recently approved CSC doctrine, Stryker Brigade Combat Teams and the Digitized Divisions along with all the maneuver brigades of the Army Transformation Divisions that are presently preparing for deployment to Iraq come with an organic MH officer and enlisted team.[24] Transformations such as this should continue.

In addition to organic mental health units, several recently developed programs utilizing NCOs as mental health assets are worthy

---

[19] Maj Richard T. Keller, interview with the author, Washington, D.C., May 21, 2003.

[20] U.S. Army Surgeon General and HQDA G1, *Operation Iraqi Freedom Mental Health Advisory Team Report,* December 16, 2003, http://www.globalsecurity.org/military/library/report/2004/mhat_report.pdf (accessed May 15, 2004).

[21] Bacon and Staudenmeijer, "A Historical Overview of Combat Stress Control Units of the U.S. Army."

[22] Marlowe interview. COL (ret.) Harry Holloway, M.D., written comments and interview with the author, Bethesda, Maryland, May 22, 2003.

[23] Holloway, written comments.

[24] Stokes written comments. Dr. Stokes also observes that the reason these teams were not organically assigned to "maneuver brigades of pre-1996 TOEs, is the consequence of Army policy, priorities and manpower constraints, not of the CSC-approved concept of operations. The maneuver brigades will continue to receive CSC team augmentation in some circumstances, while the CSC units continue to support the corps units in areas where the organic mental health teams cannot."

of greater attention. One of these programs was developed by psychiatrist Cdr Jack Pierce for the 2nd Marine Division. In this program, NCOs are taken from the operational units to serve as staff within the battalion command. Referrals from commanders for mental health care go directly to these command liaison NCOs, who interview the marine and, if evaluations or treatment are necessary, serve as case managers and coordinate care from disparate health providers.[25]

A similar program was adapted from the Royal Marines and is currently being considered by the U.S. Army.[26] This program requires one or two mid-level NCOs per company to be trained as peer mentors. They would facilitate early identification and intervention of mental health problems, help units cope with deployment related stressors, and act as liaisons between the unit and mental health services.[27] NCOs in both programs could also assist in stress education and suicide awareness training and also provide peer counseling, minimizing the need for outright evaluations. In combat, they could ensure that their units and soldiers take appropriate combat stress control steps. When acute stress casualties are incurred, they could facilitate evacuation to rear support elements or, preferably, coordinate in-unit restoration. The practice of in-unit restoration is reviewed in the next section.

NCO programs detailed here would most likely fill a critical gap. Because these NCOs are recruited from or serve in the line community, they, unlike typical mental health practitioners, would be known and trusted by line personnel and command alike.[28] Also, because they are organic to the maneuver units themselves, they would be on-hand to provide support regardless of how fast the unit

---

[25] Pierce interview.

[26] Keller interview.

[27] U.S. Army Surgeon General and HQDA G1, *Operation Iraqi Freedom Mental Health Advisory Team Report,* p. A3-6.

[28] Pierce interview.

traveled. Adoption of these programs should be considered by both the Marine Corps and Army.

Treatment Within the Unit: The Same Principles Still Apply

The unit is the first line of treatment for battle fatigue. [29] Observes Dr. J.S. Pierce, "before [marines] get to the medical facility they'd have to be way through their unit. Most marines are treated in the unit."[30] Servicemen of any branch are more open to talking to a buddy and their leaders than to personnel they don't know and who have not shared their operational experiences and hardships. Furthermore, hearing from a squad or platoon leader that their presence is critical to both the mission and the welfare of the unit is more believable and more motivating than hearing the same message from a physician.[31] When soldiers or marines are evacuated for treatment, they tend to lose contact with their unit, thus straining the very bonds that maintain their will to fight. This is particularly true during fast-moving maneuver warfare. Furthermore, it is likely that the presence of physicians, nurses, and mental health staff creates an expectation that they are ill. This seems inevitable no matter how much medical staff insist otherwise. Finally, a unit that is quick to evacuate battle-fatigued soldiers or marines may reinforce stress behaviors in other members of the unit. Consequently, handling problems within the unit, when feasible, should always be the first approach.

Signs and symptoms of battle fatigue are indeed a normal reaction to the strains of military operations, be it war or military operations other than war. Many soldiers or marines will show some mild forms of the symptoms presented in Table 5.1. This is particularly the case when individuals are exposed to battle for the first time. In

---

[29] Department of Defense, *Combat Stress.*

[30] Ibid.

[31] Ibid.

most instances, buddy and leader actions will be sufficient to stem the tide of deterioration and prevent evacuation to a medical facility.[32] Critical steps include the following:

1. Soldiers with battle fatigue symptoms should be given the assurance that their problems are normal, common reactions to immense stress. Leaders should make sure that they are assigned another soldier who can talk to them about whatever battlefield experiences or home-front stressors are troubling them. Unit medics or corpsmen, chaplains, and leaders should also be available for this purpose.
2. The expectation that the soldier will be fine must be reinforced by an authoritative source. He does not have a medical problem. Similarly, unnecessary attention should not be given to the soldier unless he is evaluated for the purposes of ruling out physical or neurological problems. Regardless, his condition should not be pitied. Sometimes, making light of the condition through joking can be helpful.
3. Leadership should, by their actions and mannerisms, demonstrate that they are calm and in control. Frightening rumors should be dispelled, and factual information on the unit's status and operational conditions should be communicated.
4. The soldier's physical needs should be replenished. If he is cold, he should be given a blanket, and if he is hot, attempts should be made to cool him down. A chance to clean up, shave, and dry off is also helpful. He should be given food and drink. To the extent that the tactical situation permits, the soldier or marine should be given an opportunity to rest and sleep. While four to

---

[32] *Battle Fatigue: Normal Common Signs What to Do for Self and Buddy* (Washington, D.C.: Headquarters, Department of the Army, GTA 21-3-004, June 1986), http://www.bragg.army.mil/528CSC/GTA21-3-4.htm (last accessed November 21, 2003), and *Battle Fatigue: Warning Signs; Leader Actions* (Washington, D.C.: Headquarters, Department of the Army, GTA 21-3-5, June 1994), http://www.bragg.army.mil/528CSC/GTA21-3-5.htm (last accessed 21 November 2003).

eight hours is preferable, even a nap of 20–30 minutes can be beneficial.

5. Finally, the soldier or marine should be given meaningful work to do, be it digging foxholes, loading ammunition, or similar simple tasks. While it is helpful for the casualty to talk through some of his problems, overemphasis on them is not helpful.[33]

These same approaches are helpful even if the symptoms of battle fatigue become more severe. However, depending on the soldier's condition, a commander may have to take additional steps. To the extent that the soldier's behavior endangers himself, the unit, or the mission, he may need to be talked down in a calm, reassuring manner. A decision should be made as to whether or not his weapon should be unloaded or taken away (although the commander should keep in mind that such actions impinge on the individual's identity as a warrior). In extreme instances, outright restraint may be necessary, although this is an option of last resort.[34]

These actions may also be necessary during combat operations or in the midst of battle. Although most battle fatigue cases occur 1–3 days after taking casualties, it is not uncommon for soldiers or marines to break down in the midst of an action. In most of these instances, the major consequence is a loss in fighting ability from the afflicted soldier. In general, the soldier will be aware enough to maintain cover from fire and, if necessary, follow the maneuvers of the unit.[35] However, there may be some instances where the soldier will expose himself or the unit to danger. These conditions may call for restraint or weapon seizure.

Clearly, treating a battle fatigue case within the unit can impose a significant burden, depending on the unit's tactical situation and the severity of the individual's battle fatigue. The unit may have been pulled from the line of contact, it may be moving in a sector near

---

[33] *Battle Fatigue: Warning Signs.*

[34] Ibid.

[35] Major Michael E. Doyle, M.D., interview with the author, Tacoma, Washington, February 20, 2003.

known enemy forces, or it may be preparing for or in the midst of a major engagement. The ongoing mission will influence a unit's holding capacity.

The severity of battle fatigue also varies. Although most cases can maintain some combat effectiveness, others in the process of recovery may not be combat effective. Among this latter group, some soldiers may require continuous observation, thus limiting the fighting availability of another soldier or marine. Others, because of their symptom presentation (erratic movement, hallucinations, aggressive behavior), may pose a danger to themselves or other members of the unit.

As such, the commander must ultimately decide what types of battle fatigue cases the unit can tolerate. One important consideration is that most cases will improve within several hours to several days if the appropriate leader actions are taken. This is important given that there is no guarantee that the medical holding facility will be able to return the soldier to duty. Consequently, within-unit restoration of a battle-fatigued serviceman may be worth the short-term inconvenience of having to care for him. Another available option is to temporarily transfer the soldier to a combat service support (CSS) unit operating in the area where the soldier or marine can be given an opportunity to rest and also perform meaningful work in a less hazardous environment. The soldier can return to the unit when the tactical situation permits.

Ultimately, there will be cases that require evacuation for medical evaluation. This will be necessary for cases that suffer from gross behavioral disturbances such as hallucinations, suicidal behavior, or erratic behavior, as well as cases that have lost combat effectiveness and have failed to improve after attempts at in-unit restoration.

Under these circumstances, unit members and commanders should still apply the general principles of expectancy and reassurance. They should tell the soldier that he is a vital member of the team and that they need him and look forward to him returning to the unit. Furthermore, if possible, members of the unit should visit him in the CSC's restoration unit or the CSS unit if he is labeled a "hold" casualty.[36]

While educational documents such as this monograph may be helpful in educating members of the line community about stress reactions, it is important to observe the critical educational role played by mental health staff organic to divisions or brigades. Members of these staffs both train and mentor unit members about CSR. They are also specially trained to screen soldiers presenting with CSR for other physical or mental health problems that may require other treatment or evacuation. A key function of mental health professionals is extending this training to medical personnel, chaplains, and on to individuals within the units.[37]

Urban-Specific Applications

Evidence from both the Battle of Hue and the Battle of Lebanon suggest that increased concentration of forces during a relatively slow-moving urban battle may actually facilitate treatment for battle fatigue. Dr. Arieh Shalev, a former physician with the IDF, observed:

> One characteristic of fighting on urban [terrain] is the proximity of [the] first aid station [to the line of contact] and the immediate evacuation [of wounded] to facilities. When units get organized in urban operations it is clear that distance between shooting and first aid station is often minimal vs. armored [conflicts] where the distance [is greater].[38]

Dr. Shalev goes on to note that the end result of such proximity was an increased ability to return soldiers to duty: "In Beirut, the front was not mobile, people [with CSR] could return to their troops vs. open terrain [where soldiers with CSR] lose their units and it's

[36] Department of Defense, *Leaders' Manual for Combat Stress Control.*

[37] Stokes, written comments.

[38] Arieh Shalev, M.D., interview with the author, Jerusalem, Israel, July 7, 2003.

harder to return [soldiers] to duty."[39] Similarly, Dr. J. Price Brock, the battalion surgeon for 1/5 marines during the Battle of Hue, noted that he treated two battle fatigue casualties just 12–15 blocks from the main fighting. Under these types of conditions, returning battle-fatigued casualties back to duty requires at worst a long walk.[40] In addition, although some problems persist, CSC and division mental health sections during stability operations in Iraq reported return-to-duty rates that averaged 96 percent.[41] Of course, these types of conditions do not always apply. During Operation Just Cause, in which groups of U.S. soldiers were dispersed throughout the city, the required evacuation distance was much larger.

Although the proximity between the line of contact and forward psychiatric treatment centers increases the speed of evacuation, the stress casualty, according to Dr. Ron Levy, the chief of psychology for the IDF during the Lebanon War, will still require assistance in his trek to the rear.

> As long as [the evacuation] is done with buddies and field commanders. Just like you get a bullet and can still function, in leg or arm, you have to retreat. Who helps you to do that? Your buddies. That is the thing that has to be done, "Sgt take your buddy whether he has a bullet in his arm or leg or has a combat reaction, take him backwards, slowly be careful don't lose sight of where you're going, take your gun." So I call it, when I speak, "temporary ineffectiveness," so you can get a bullet [in the arm or leg] and can still function.[42]

At times, however, circumstances, especially during high-intensity operations, may severely restrict the ability of units to evacuate their wounded. Tactical units may get cut off from their rear

---

[39] Ibid.

[40] LCDR (ret.) J. Price Brock, M.D., interview with the author, Abilene, Texas, March 6, 2003.

[41] U.S. Army Surgeon General and HQDA G1, *Operation Iraqi Freedom Mental Health Advisory Team Report.*

[42] Levy interview.

support, or irregular combatants might improperly target ambulances. Under these high-threat circumstances, it is noteworthy that the occurrence of stress casualties is often greatest following the combat scenario. According to Dr. David Marlowe, "When engaged in life and death struggle is not when you get the casualties [it is afterward]. Look at the scenario in [the film] *Black Hawk Down*. There is no time to get stress casualties. You do what you're doing or you're dead."[43]

When stress casualties do occur, members of tactical units should be prepared. Under these circumstances, the soldier with CSR may have to be held at the line of contact. Individuals interviewed for this report generally suggest steps akin to the in-unit restoration efforts described above.[44] Dr. Levy does so as well. He notes specifically that soldiers with battle fatigue will need to be placed in a setting that can best provide them a sense of security, be it "a basement in an urban area or deep in a trench." He goes on to suggest:

> So recuperation time and ambience and that has to be done according to local understanding of medics and with the embrace and the hug. Using both physical embrace and the hug and the right words. And then in every language "take it easy," the local verbs and nouns or jokes, to get the guy back in the right mood. . . . And especially the person who is supposed to take care of him. Say you are a company commander and you need all the combatants you have [and there are] 2–3 combat reactions. You take the guy who is half in combat shock to take care of them. He will do a bad job so you have to take the good guy, who has some experience, a little older and not necessarily the medic. It has to do with the personality more [than] the medical training. He is a combatant, you send him back there to speak in their language, get him back. Give them back that sense of being part of us. . . "Stay with them for few hours and come back and report. Are you telling them jokes, are they laughing, are they eating or sleeping?" What you get are three combatants back in

[43] Marlowe interview.

[44] COL Robert Collyers (Australian Army), interview with the author, Brisbane, Australia, March 7, 2003.

24 hours. And that can be significant in your ability to fight the mission.[45]

The Success of PIES

Hell, I guess somebody's got to fight this god-damned war.

—*Statement overheard from a soldier departing from a forward psychiatric treatment center in the Korean War*[46]

The goal of forward psychiatry or combat stress control is to return soldiers to their fighting units and limit the tide of psychologically precipitated evacuations. Commanders and senior NCOs should be aware of its efficacy. The success of forward psychiatric interventions (PIES) as administered by mental health assets separate from the soldier's original unit in a relatively stable battle space is reviewed below, along with the risk of relapse in treated soldiers. In World War II, published return-to-duty (RTD) rates varied from 50 to 90 percent, successful outcomes by most any measure. However, the personal records of many psychiatrists depicted more limited success, with battle-fatigued soldiers being returned to their original combat units at rates of 22–60 percent. In addition, in at least one of these instances relapse rates reached 43 percent. The discrepancy between published and personal records has been partly attributed to psychiatrists attempting to enhance their reputations within the Army and an attempt to maintain the morale of the troops.[47]

---

[45] Levy interview.

[46] W. L. White, *Back Down the Ridge* (New York: Harcourt, Brace and Company, 1953), referenced in Ritchie, "Psychiatry in the Korean War," p. 900.

[47] Reviewed in Jones and Wessely, "Forward Psychiatry in the Military." These observations applied to World War II–era psychiatrists. It is unclear whether similar pressures applied to RTD rates in subsequent military campaigns.

Two additional World War II studies have sought to more fully document the efficacy of forward psychiatry. Albert Glass, for example, surveyed 393 U.S. troops who were committed to the Apennines campaign between March and April 1945. Here, 54 percent of stress casualties treated in a division-level treatment unit were returned to duty, but only 30 percent were actually sent back to combat units.[48] Yet, in the words of Dr. Jones, who reviewed this study,

> two-thirds of those who later relapsed did so by other routes (principally disease, injury or military offence), while 25% of those returned to combat units and then found to be ineffective were retained by their commanders . . . . Glass concluded that it was feasible to return the vast majority of neuropsychiatric casualties to non-combatant base or support duties, but only 30% to active duty.[49]

Alternatively, another World War II U.S. study followed 316 battle fatigue casualties for 1–3 months after they were returned to full combat duties. All casualties were treated at two forward psychiatric treatment centers operating in the Seventh Army. These treatment centers were probably second-echelon facilities and thus treated soldiers who failed to adequately recover at division. Follow-ups were conducted via letters to the soldier's commanding officer that requested a summary of the former patient's disposition. The 316 follow-up number reflects the 90 percent follow-up rate for whom letters were returned. Results demonstrated that only 26.6 percent of this sample were reported to be present in their unit and performing reasonably well, in contrast to nearly 70 percent who were no longer present for duty. Of those no longer present, 48 percent were either readmitted for battle fatigue or were absent through other routes such

---

[48] A. J. Glass, "Effectiveness of Forward Neuropsychiatric Treatment," *Bulletin of the U.S. Army Medical Department,* Vol. 7 (1947), pp. 1034–1041, referenced in Jones and Wessely, "Forward Psychiatry in the Military."

[49] Jones and Wessely, "Forward Psychiatry in the Military," p. 9.

as absent without leave (AWOL), self-inflicted wounds (SIW), or being administratively discharged or reassigned to CSS.[50]

In the Korean War, a three-echelon system of forward psychiatric care was instituted only two months after the onset of hostilities. The three echelons included the division, theater (Korea), and the zone of interior (Japan and CONUS). At the divisional level, 50–70 percent of treated exhaustion cases were reportedly returned to their original combat units. Relapse rates, unfortunately, were not provided.[51]

The most extensive report detailing the relationship between RTD and forward psychiatric treatment stems from the 1982 Lebanon War. This retrospective study evaluated the contributions of proximity, immediacy, and expectancy to RTD rates and long-term psychological trauma (PTSD). The authors studied soldiers with combat stress reactions treated at varying distances from the front line (proximity), those whose treatments ranged from immediate to two or more days after onset of symptoms (immediacy), and those given expectancies that ranged from returning to duty at all costs to unclear or ambiguous expectations (expectancy). The authors reported that the more these principles were applied (from zero to all three treatment principles), the greater was the percentage returned to duty (22 percent if none of the principles were applied to 60 percent if all three were) and the lower the percentage reporting PTSD one year later (ranges from 71 percent if none of the principles were applied to 40 percent if all three were present in treatment).[52]

At face value, this study validates the use of PIE principles in forward psychiatry. Yet scientific limitations may weaken this interpretation with regard to the development of PTSD symptoms. As

---

[50] A. O. Ludwig and S. W. Ranson, "A Statistical Follow-Up of Effectiveness of Treatment of Combat-Induced Psychiatric Casualties: I. Returns to Full Combat," *The Military Surgeon*, January 1947, pp. 51–64.

[51] A. J. Glass, "Psychiatry in the Korean Campaign," *United States Army Medical Bulletin*, Vol. 4, No. 10 (1953), pp. 1387–1401.

[52] Solomon and Benbenishty, "The Role of Proximity."

suggested by Jones and Wessely,[53] the authors note that a given soldier's prognosis or expected treatment outcome influenced whether or not he was returned to duty and that returning the soldier to duty may have positively influenced outcome. Thus, subsequent rates of PTSD that are related to RTD rates may simply be a reflection of the soldier's original battlefield prognosis. Moreover, it may be that prognosis directly influenced the level of expectation a soldier was given regarding returning to duty (i.e., soldiers given greater expectations precisely because their prognosis was good). The presence of a positive prognosis may also have influenced evacuation to a front-line treatment center. Thus, it remains unclear whether expectation and front-line treatment directly influenced RTD rates or whether RTD was mediated simply by real and perceived prognosis. While the study shows that a significant proportion of soldiers were returned to duty, the individual contribution of each of the PIE elements remains unclear.

Regardless of the methodological issues addressed for the Solomon and Benbenishty study, it does seem to demonstrate one very important point: individuals with a past history of a CSR do have relatively high rates of PTSD. The long-term outcome of Israeli CSR casualties supports this conclusion.[54] For example, a small sample of decorated heroes (n=98), CSR casualties (n=112), and combatant controls (n=189) were questioned about PTSD symptoms approximately two decades following the Yom Kippur War.[55] The response rate for this survey ranged from 66 to 74 percent. As seen in Figure 5.2, individuals with a documented CSR were more likely to self-report symptoms relevant to both past and present PTSD than either the groups of controls or decorated soldiers.[56]

---

[53] Jones and Wessely, "Forward Psychiatry in the Military."

[54] Solomon, Benbenishty, and Mikulincer, "A Follow-Up of Israeli Casualties of Combat Stress Reaction."

[55] R. Dekel, Z. Solomon, et al., "Combat Exposure, Wartime Performance, and Long-Term Adjustment Among Combatants," *Military Psychology,* Vol. 15 (2003), pp. 117–131.

[56] The retrospective nature of these data, the small sample size, and the use of control subjects who were not randomly sampled and were not chosen from the CSR's original units

Summary of Treatment Outcome

The goal of forward psychiatry is to return battle-fatigued servicemen to duty, thus limiting the number of individuals evacuated out of theater. This goal has met with some success. Soldiers treated near the front lines in psychiatric treatment centers are more likely to be returned to duty. World War II–era RTD rates vary widely from 22 to 80 percent, with some casualties requiring reassignment to support units. RTD rate variations most likely depend on a number of fac-

Figure 5.2
The Presence of Past and Current PTSD in a Sample of Israeli Yom Kippur War Veterans

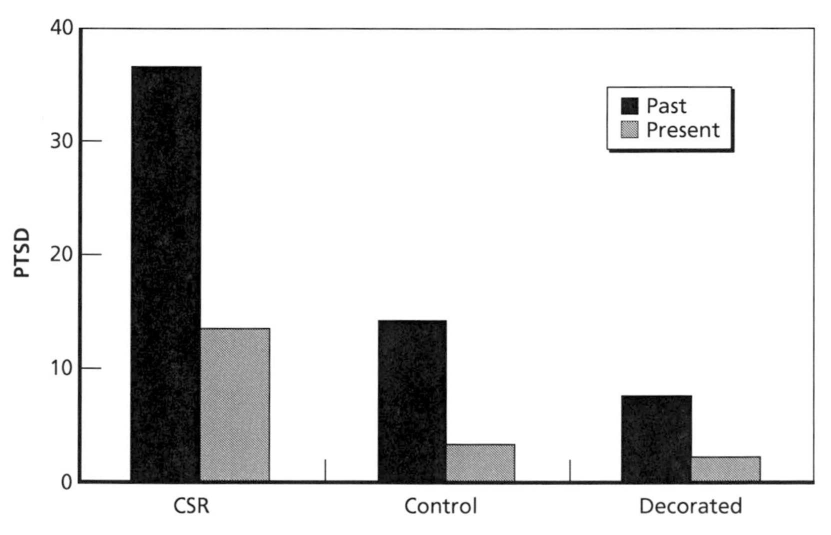

RAND *MG191-5.2*

SOURCE: Dekel et al., "Combat Exposure, Wartime Performance, and Long-Term Adjustment Among Combatants," p. 124.

(thus controlling for combat exposure) limit the extent to which the group's relative risk of PTSD can be taken literally. The data do suggest, however, that CSR casualties have an increased risk for PTSD.

tors, including the severity of battle fatigue symptoms and the hospital's logistical ability to return soldiers to their original units. This latter factor may be a problem during military actions involving fast-moving battlefields like that seen in the Gulf War. It may be possible to address this potential issue by providing more treatment within combat units or combat service support units and through the assistance of NCO peer mentors. However, the success of in-unit restoration, with or without peer mentors, has, to the knowledge of these authors, never been evaluated. It should be a focus of future field investigations. It is clear that some soldiers and marines relapse after being treated, but the extent of this risk of relapse is not well known. CSR also increases one's risk of a subsequent diagnosis of PTSD. It currently remains unknown whether successfully administered forward psychiatric treatment can reduce the risk of PTSD.

CHAPTER SIX

Preventing the Fall: Suggested Steps Toward the Prevention of Stress Reaction

During highly intense combat operations, rates of stress casualties have been known to nearly equal the number of physical casualties in certain units.[1] Across an entire theater, rates approximating 25 percent of physical casualties are common. While these rates are high, evidence suggests that there is great variability in stress reaction rates between combat units. This is exemplified in the fact that airborne units during World War II reported stress reaction rates nearly a fifth of those reported by regular infantry divisions, despite experiencing combat of equal or greater intensity.[2]

Data such as these demonstrate that the risk of stress casualties is a modifiable one. Commanders and NCOs can take a number of common-sense steps that may significantly reduce the risk of stress casualties.[3] These steps also carry the added benefit of improving combat effectiveness. The recommendations included in this monograph are largely based on interviews with combat soldiers and marines and medical authorities. They are organized in terms of pre- and

[1] Belenky, Tyner, and Sodetz, *Israeli Battle Shock Casualties.*

[2] Hessin, "Neuropsychiatry in Airborne Divisions."

[3] Ideally, the interventions we propose for reducing the risk of stress casualties would be backed by prospective scientific research. This is unfortunately not the case, as the battlefield poses clear limits to careful scientific observation. Recommendations are thus based on best-known risk factors and expert opinion.

postdeployment efforts, and, where warranted, they are tailored to the demands of urban operations.

Predeployment

Screening

> If screening [were] to weed out anyone who might develop a psychiatric disorder, it would be necessary to weed out everyone.
>
> —*John W. Appel, M.D.*[4]

> [You] can't predict . . . he's the guy . . . sometimes an individual acts and [his] personality is not reflective of what will happen when it gets down to the nitty gritty.
>
> —*Col (ret.) Myron Harrington*[5]

During World War II, authorities fervently sought to prevent psychological breakdown through a massive screening program. As noted early in this report, this program proved a massive failure, severely restricting the number of military inductees while failing to prevent psychological breakdown in combat. Despite this poignant historical lesson, some military authorities still believe that such a program could work in the near future.

Unfortunately, a number of factors collude to severely limit the success of any screening program designed to prevent the onset of battle fatigue. It remains unclear what individual characteristics that

---

[4] J. W. Appel, "Preventive Psychiatry," in R. S. Anderson, A. J. Glass, and R. J. Bernucci (eds.), *Medical Department, United States, Neuropsychiatry in World War II*, Vol. I, *Zone of the Interior* (Washington, D.C.: U.S. Government Printing Office, 1966), pp. 373–414, 391.

[5] Col (ret.) Myron Harrington, interview with the author, Mount Pleasant, South Carolina, February 24, 2003.

could be known at the time of recruit induction or a military assignment actually predict vulnerability to breakdown. While investigators have identified several variables associated with breakdown or stress tolerance, no single variable has shown a sufficient amount of consistency to prove useful in screening. Furthermore, even if such a factor or, more likely, constellation of factors were identified, it is highly unlikely that researchers could design a paper-and-pencil test or interview that could accurately measure them. Measurement accuracy is critical for screening programs in that a given test must be able to accurately predict those who will have a given problem, such as CSR (referred to as sensitivity), and it must accurately predict which individuals will not get CSR (referred to as specificity). Present-day test construction does not permit sufficient levels of sensitivity or specificity to warrant a full-scale screening program aimed at the prevention of battle fatigue.[6]

To illustrate this point, in World War II, 2,054 men who were rejected by the Selective Service were subsequently inducted into the Army. Although 18 percent of this sample was ultimately discharged due to psychiatric causes, the remaining 82 percent were given a satisfactory duty rating. This is in contrast to a 94 percent satisfactory rating for all enlisted personnel. The authors concluded that 1,992,950 soldiers were unnecessarily prevented from serving their country.[7] Several studies also show that the presence of psychological disorders during training does not necessarily predict the occurrence of stress casualties.[8]

---

[6] E. Jones, K. C. Hyams, and S. Wessely, "Screening for Vulnerability to Psychological Disorders in the Military: An Historical Survey," *Journal of Medical Screening,* Vol. 10 (2003), pp. 40–46. Dr. Jones also points out in this review that screening programs have proved successful or promising for other important factors such as intelligence or psychosis, and are being used in regard to trade deployment, officer selection, and likelihood of completing of boot camp.

[7] J. R. Egan, L. Jackson, and R. H. Eanes, "A Study of Neuropsychiatric Rejectees," *Journal of the American Medical Association,* Vol. 145 (1951), pp. 466–469.

[8] M. R. Plesset, "Psychoneurotics in Combat," *American Journal of Psychiatry,* Vol. 103 (1946), pp. 87–88.

Finally, it is worth noting that screening programs may have many adverse consequences.[9] In addition to their high financial costs,[10] they would also label individuals as having a mental illness (or a greater propensity for one), a stigma that they might carry all their lives. It has also been suggested that a screening program would generate a degree of mistrust among the general military population if they feared that honest responses in health screenings would be used against them.[11] Future advances in psychiatry and test construction notwithstanding, it seems worthwhile for the military to focus on those matters that have greater impact on breakdown during service and address them rather than attempting to screen candidates with respect to propensity for CSR. These include training, morale, cohesion, and a number of others, and they are the focus of the remainder of this monograph.

Soldier Indoctrination

Military leaders should attempt to build a sense of unit pride and camaraderie when any soldier joins a unit. According to psychiatrist Michael Doyle, a command must introduce "new recruits to the culture, help sponsor them into the unit so they know the unit history, [know its] motto and what it means, [and understand the] lore of the unit—what battles is it famous for."[12] Gregory Fontenot went to great lengths to expose new recruits to his 2nd Battalion, 34th Regiment's culture prior to 1991 combat in Iraq, including devoting a section of his unit's conference room to the history of the organization. New soldiers could go there to observe unit artifacts, photo-

---

[9] As reviewed in Jones, Hyams, and Wessely, "Screening for Vulnerability."

[10] R. J. Ursano, T. A. Grieger, and J. E. McCaroll, "Prevention of Posttraumatic Stress: Consultation, Training, and Early Treatment," in B. A. van der Kolk, A. C. McFarlane and L. Weisaeth (eds.), *Traumatic Stress: The Effects of Overwhelming Experience on Mind, Body and Society* (New York: The Guilford Press, 1996).

[11] C. E. French, R. J. Rona, et al., "Screening for Physical and Psychological Illness in the British Armed Forces: II. Barriers to Screening—Learning from the Opinions of Service Personnel." *Journal of Medical Screening,* Vol. 11, No. 3 (2004), pp. 153–161.

[12] Doyle interview.

graphs, and a slide and videotape show featuring unit history and base information.[13] According to Doyle, orientations such as this build loyalty to the unit, reduce a soldier's sense of newness, and enhance cohesion.[14] According to a World War II study, soldiers with pride in their unit are more likely to state that they are ready for combat.[15] The author of this study reasoned that

> pride in outfit for the combat man included something over and above personal identification with the "other guys" and the leaders in the outfit. He took pride in its history as well as its present, and identified with the men who had died in the outfit as well as the living. As it has been suggested, he owed it to them—they hadn't got off easy.[16]

Training

> "Did you see those Japanese firing at us?" he screamed to the guy next to him. "No," the leatherneck answered, deadpan. "Did you shoot them?" "Gee, no," Buchanan replied, "That didn't occur to me. I've never been shot at before."
>
> —*Conversation between two marines, D-Day, Iwo-Jima*[17]

As a matter of background, stress is known to have a number of deleterious consequences in both human performance and mental health. On a performance level, studies have shown that stress impairs many behaviors that are critical to effective combat performance. Such behaviors include marksmanship, decisionmaking, and teamwork, to name but a few.[18] The reasons for impairment are varied. First, stress

---

[13] G. Fontenot, "Fear God and Dreadnought: Preparing a Unit for Confronting Fear" *Military Review* (July–August, 1995), pp. 13–24.

[14] Doyle interview.

[15] Stouffer et al., *The American Soldier*.

[16] Ibid., p. 139.

[17] J. Bradley and R. Powers, *Flags of Our Fathers* (New York: Bantam Books, 2000), p. 156.

[18] Watson, *War on the Mind*. J. E. Driskell and J. H. Johnston, "Stress Exposure Training," in J. A. Cannon-Bowers and E. Salas (eds.), *Making Decisions Under Stress: Implications for*

seems to disrupt goal-oriented cognitions or thoughts. A soldier who must split his attention between overwhelming levels of anxiety and the highly consuming task of clearing a building of enemy combatants will probably perform the latter task to a substandard level.[19] The combat environment is replete with novel and frightening stimuli such as dead civilians or wounded comrades. These stimuli may distract soldiers from their critical mission.[20] Stress can also lead to a loss of confidence. It seems self-evident that a soldier who loses faith in his ability to successfully perform a task will lose a critical edge. Research shows that self-efficacy or the belief that success is inevitable improves levels of performance.[21]

Beyond performance, stress increases the risk of psychological breakdown. It is understood in psychiatry that negative thoughts or physiological symptoms of fear, left unchecked, can in and of themselves be a cause of increased anxiety.[22] Referred to as "the downward spiral of anxiety," thoughts of anxiety lead to impaired performance (or at least the interpretation of poor performance) that in turn breeds greater levels of fear. The cycle self-perpetuates to a point where anxiety becomes overwhelming. Similarly, we know that a lack of confidence is a risk factor for breakdown.[23] A soldier who loses confidence in himself and his leadership becomes vulnerable to battle fatigue. As a result of these factors, it is critical that military training

---

Individual and Team Training (Washington, D.C.: American Psychological Association, 1998), pp. 191–217.

[19] B. P. Lewis and D. E. Linder, "Thinking About Choking? Attentional Processes and Paradoxical Performance," *Personality and Social Psychology Bulletin,* Vol. 23 (1997), pp. 937–944.

[20] E. P. Lorch, D. R. Anderson, and A. D. Well, "Effects of Irrelevant Information on Speeded Classification Tasks: Interference Is Reduced by Habituation," *Journal of Experimental Psychology: Human Perception and Performance,* Vol. 10 (1984), pp. 850–864. D. J. Madden, "Aging and Distraction by Highly Familiar Stimuli During Visual Search," *Developmental Psychology,* Vol. 19 (1983), pp. 499–507.

[21] A. Bandura, "Self-Efficacy: Toward a Unifying Theory of Behavioral Change," *Psychological Review,* Vol. 84, No. 2 (1977), pp. 191–215.

[22] R. M. Rapee and R. G. Heimberg, "A Cognitive-Behavioral Model of Anxiety in Social Phobia," *Behaviour Research and Therapy,* Vol. 35, No. 8 (1997), pp. 741–756.

[23] Stouffer et al., *The American Soldier.*

programs help soldiers overcome the negative consequences brought about by stress.

Stress Exposure Training. Stress exposure training (SET) provides a framework in which fear can be attenuated and tasks can be trained such that performance withstands the aversive effects of stress. This program, based on medical, psychological, and military research and articulated by James Driskell and Joan Johnston,[24] involves a three-step approach of (1) information provision, (2) skills acquisition, and (3) application and practice with the goal of confidence building. Each of these phases is briefly described below.

Phase 1: Providing Knowledge of the Stress Environment. Research has demonstrated that individuals given information about an impending stressful event experience a reduction in anxiety and an increase in both confidence and objectively measured levels of performance.[25] It has been hypothesized that preparatory information renders a given situation more predictable, thereby limiting cognitive resources being applied to guesswork. Moreover, it can render many aspects of the stressful environment less novel and therefore less distracting.[26]

Accordingly, the first phase of SET requires that soldiers learn about how they are likely to feel during combat (*Sensory Information)* and about the actual stressors inherent in the combat environment and how those stressors affect performance (*Procedural Information).*

For sensory information, servicemen and women must be taught the full range of emotional symptoms (e.g., anxiety, confusion) and physical signs (e.g., heart palpitations, release of bladder contents) of stress, emphasizing that such experiences are normal upon first exposure to combat. As a result, soldiers will be less likely to overinterpret normal anxiety symptoms as signs of impending breakdown, which otherwise might serve to further increase anxiety. Such training will

---

[24] Driskell and Johnston, "Stress Exposure Training."

[25] C. M. Inzana, J. E. Driskell, et al., "Effects of Prepatory Information on Enhancing Performance Under Stress," *Journal of Applied Psychology,* Vol. 81 (1996), pp. 429–435.

[26] Driskell and Johnston, "Stress Exposure Training."

also tend to limit the extent to which such symptoms distract attention away from the combat mission.

The particular stressors inherent in the combat environment are numerous. In addition to dead and wounded comrades, stressors particular to the urban combat environment include noncombatant casualties, exhaustion secondary to clearing multifloor buildings, and multiple fields of enemy fire, to name but a few. These stress-provoking stimuli should be openly discussed and subsequently modeled in realistic training exercises. In addition, the effects that these stressors have on soldier performance, such as marksmanship and teamwork, should be thoroughly reviewed.

Finally, instrumental information gives the soldier the means to counteract the negative consequences of stress. Suggestions might include not looking at the faces of noncombatant dead or using deep-breathing exercises to limit anxiety. In addition, soldiers should be reassured that adhering to tactical training principles will increase their chances of surviving unscathed.

Phase 2: Skills Acquisition. There are multiple strategies that can be communicated to the soldier or marine to improve stress resiliency. Cognitive control strategies can help soldiers learn to recognize task-irrelevant thoughts and replace them with task-relevant thoughts. For example, soldiers should be reminded that worrying about remaining time during a timed marksmanship task only impairs their marksmanship and that thinking about the plight of soldiers on a higher floor only distracts them from clearing their own assigned room. Developing such skills not only improves task performance but also reduces anxiety.[27]

In addition, soldiers can be taught time-sharing skills. For example, marksmanship learned under calm conditions may become impaired in the presence of noise or moving stimuli, two distractions that are omnipresent in the urban environment. However, pairing marksmanship with these distractions during training should improve

---

[27] J. Wine, "Test Anxiety and Direction of Attention," *Psychological Bulletin*, Vol. 76 (1971), pp. 92–104. B. A. Thyer et al., "In Vivo Distraction—Coping in the Treatment of Test Anxiety," *Journal of Clinical Psychology*, Vol. 37 (1981), pp. 754–764.

task performance.[28] In addition, combat often requires multiple tasks to be performed at the same time. A soldier may have to engage one target while also scanning the environment for additional threats. Data suggest that training tasks individually may still result in impaired performance when the tasks are combined.[29] However, significant improvements in multitask performance can be obtained when the tasks are paired during training.[30]

Deliberate over-learning is another important tool in developing skills that are resilient to the impairing effects of stress.[31] Over-learning simply refers to repetitive drill above and beyond what is required to develop an initial skill. Driskell and Johnston suggest that over-learning should take place under conditions similar to the anticipated stressful conditions.[32]

Finally, Driskell and Johnston suggest that team skills should be a focus of training. They note that group coordination deteriorates under stressful conditions, most likely due to a narrowing of attention in individual group members.[33] In a separate study, Driskell and colleagues showed that teamwork under stressful conditions could be improved through an intervention that provided information on teamwork and the deleterious consequences of stress, trained teamwork skills, and required practice of teamwork under stressful conditions.[34]

Phase 3: Confidence Building Through Application and Practice. Tasks learned under nonstressful conditions may deteriorate when the

---

[28] R. N. Singer, J. H. Cauraugh, et al., "Attention and Distractors: Considerations for Enhancing Sport Performances," *International Journal of Sport Psychology,* Vol. 22 (1991), pp. 95–114.

[29] Driskell and Johnston, "Stress Exposure Training."

[30] R. F. Gabriel and A. A. Burrows, "Improving Time-Sharing Performance of Pilots Through Training," *Human Factors,* Vol. 10 (1968), pp. 33–40.

[31] J. E. Driskell, R. P. Willis, and C. Cooper, "Effect of Overlearning on Retention," *Journal of Applied Psychology,* Vol. 77 (1992), pp. 615–622.

[32] Driskell and Johnston, "Stress Exposure Training."

[33] Ibid.

[34] Ibid.

same task is performed under duress. Training during stressful situations limits the extent of this deterioration. According to Driskell and Johnston,[35] training under conditions that approximate the stressors of the operational setting serves to improve overall performance and confers experience with performance problems engendered by the stressor. Not only will stress exposure training improve performance during battle conditions, but it should also increase the soldier's confidence, a critical step in limiting the extent of emotional breakdown.

One common criticism of stress exposure training for military personnel is that the stress of combat can never be accurately and fully characterized in training exercises. While this is true, research does suggest that complete and accurate re-creation of operational stressors is not necessary. For example, performance improvements derived from training under one stressor (e.g., noise) will generalize to other stressors (e.g., time pressure)[36] and that improvements gained for one task will generalize to another.[37] Another criticism is that stress exposure may itself produce such a degree of stress that soldiers become even more fearful than before the training. This may especially happen when an extremely potent stressor is exposed en masse to soldiers who have no effective means of dealing with it. To avoid this problem, stressors must be introduced incrementally so as to facilitate habituation and to allow soldiers the opportunity to develop necessary coping skills.

Initial learning for task performance best takes place under nonstressful environments. Once learned, however, training should take place under increasingly stressful environments culminating in relatively complex and accurate combat exercises. In addition, tasks should not be introduced that are virtually impossible to master. In-

---

[35] Ibid.

[36] G. Vossel and L. Laux, "The Impact of Stress Experience on Heart Rate and Task Performance in the Presence of a Novel Stressor," *Biological Psychology,* Vol. 6, No. 3 (1978), pp. 193–201.

[37] J. E. Driskell, J. H. Johnston, and E. Salas, "Does Stress Training Generalize to Novel Settings?" *Human Factors,* Vol. 43 (2001), pp. 99–110.

stead, ensuring initial task accomplishment builds confidence. As the training program progresses, task difficulty can be increased.

Developing Urban-Specific Skills and Recreating the Combat Environment. A number of specific skills must be taught to soldiers preparing for urban operations. A sample of requisite skills is reviewed below for three military occupational specialties (MOSs).

Infantrymen clearly play a prominent role in any combat environment, and the urban environment is no exception. The most obvious skill for this MOS is marksmanship and the ability to rapidly sight and engage targets. Also necessary is the ability to move through buildings and across streets heavily exposed to gunfire. The effective clearance of rooms of enemy combatants in multistory buildings is a must, and given the likelihood that noncombatants will also be encountered, soldiers must have the ability to discriminate before pulling the trigger. In addition, city centers, with a vast array of structures and potentially confusing layouts, especially in areas where landmarks are few, call for highly developed navigation skills. Blockage of line-of-sight communication systems caused by multistory buildings necessitates training with intermittent communication between leaders and those who are led. It is often said that urban combat is a small unit leader's fight. Lieutenants and NCOs should be trained accordingly and be given experience in tactical decisionmaking.

Support units should be given similar urban-specific skill training. For the engineer, necessary skills include removal of street and building entrance obstacles and clearance of mines and booby traps, many of which will be improvised. Following active combat operations, cities will require rapid restoration of services such as electricity and water. As seen in the early stages of stability operations for Operation Iraqi Freedom, many of these services will be subject to repeated sabotage that must be remedied in short order.

Soldiers operating as part of supply and transport units should develop a skill set unique to their own provision of service. For these units, city navigational skills are important and will at times require circumvention around numerous types of obstacles. Given the risk of ambushes, these soldiers should maintain effective combat skills so that attacks can be aggressively repelled. Supply and transport soldiers should also be skilled at handling anxious and angry crowds of civilians at food and supply distribution points and be able to effectively transition to this duty from combat engagements.

When training for Military Operations on Urbanized Terrain, a number of urban-specific stressors should be incorporated into the training environment. As noted before, the presence of civilians on the battlefield presents a unique stressor to the soldier trained to value noncombatant life. The fact that many of these civilians could take an active combatant role will do little to ease the conscience of U.S. servicemen who are forced to take their lives. Dr. Ron Levy, the head of IDF psychology during the Lebanon War, notes that soldiers must be reconditioned to see images of civilian death and destruction.[38] Actors, taking the role of civilians during urban training exercises, should be liberally employed. They should not only portray fearful civilians requiring active manipulation by military personnel, but also combatant civilians who have to be targeted and engaged and civilian bodies on the battlefield. In addition, the visual scene of dead and wounded too often commonplace on an urban battleground should be replicated through the use of mulage kits.[39] Soldiers and marines should also develop an understanding that critically wounded soldiers may not have ready access to lifesaving medical support. To this end, soldiers ought to play the role of wounded comrades dying on a street corner exposed to withering gunfire or, as the film *Black Hawk Down*[40] vividly portrays, a soldier fighting for life without access to needed medical evacuation. The element of surprise can also be introduced into training exercises. LtCol (ret.) Robert Barrow notes that Situational Training Exercises (STE), in which marines are briefed for one situation only to confront an entirely unexpected situation, are especially helpful.[41] He further suggests that real-life

---

[38] Levy interview.

[39] Mulage kits refer to placing animal innards or the like on a "wounded" soldier's body to simulate an actual battle wound.

[40] *Black Hawk Down*, directed by Ridley Scott, Columbia Pictures, 2001.

[41] Barrow interview.

situations derived from past operations be utilized to emphasize to soldiers and marines that such scenarios do in fact take place.

Simunitions (chalk or similar ammunition fired from actual combat weapons) are another important tool in recreating the stress of combat. LtCol (ret.) John Allison states that simunitions are a "great training tool for the urban environment." With them, "people now duck below the window and don't want to get hit."[42] They also learn the consequences of poor battlefield performance.

Finally, live-fire exercises with actual ammunition are an important element in training. Referring to the October 3, 1993 TFR raid in Mogadishu, SFC Matthew Eversmann stated his belief that

> most would agree that the confidence each Ranger had under fire was proportional to the amount of live firing he had done. Shooting in close proximity builds confidence and cohesion as does marksmanship training. Being used to live firing and weapons handling in all types of situations was invaluable. If there is one thing to train on, or one tool to use, it has to be live fire.[43]

Whether training programs are geared toward developing urban-specific capabilities or capabilities for operations on other types of terrain, it is critical that commanders determine situation-specific skill sets and anticipate relevant environmental and operational stressors. Once those skills are trained, they must be honed in the midst of increasingly realistic and stressful environments.

Physical Fitness. Physical fitness is valuable for soldiers who plan to enter any combat environment, but especially those confronting urban combat. LtGen (ret.) George R. Christmas argues that "those [who are] physically fit perform better in [the] urban environment."[44]

---

[42] Allison interview.

[43] M. Eversmann, "The Urban Area During Support Missions Case Study: Mogadishu, The Tactical Level I," in R. W. Glenn (ed.), *Capital Preservation: Preparing for Urban Operations in the Twenty-First Century; Proceedings of the RAND Arroyo-TRADOC-MCWL-OSD Urban Operations Conference, March 22–23, 2000* (Santa Monica, CA: RAND Corporation, CF-162-A, 2001).

[44] LtGen (ret.) George Christmas, interview with the author, Stafford, Virginia, February 28, 2003.

CSM Michael Hall and SFC Michael Kennedy, U.S. Army Rangers speaking about a three-week program of increased physical fitness training incorporated by the Rangers, bolstered this insight. This program not only resulted in increased physical strength, but also had the added benefits of reduced injuries, increased success of executing combat tasks, and lessened mental exhaustion.[45] Importantly, to the extent that physical fatigue increases the risk for stress reactions, a physical training (PT) program that increases strength and endurance should increase mental resiliency. Like any training program, PT workouts should be partially geared to the anticipated physical demands. For example, maneuvering through urban terrain often requires sprints through fields of enemy fire. PT programs should not only develop speed but also foster skills of maintaining environmental awareness while a soldier is maneuvering.

Cohesion

> Generally speaking, the need for military virtues becomes greater the more the theater of operations and other factors tend to complicate the war and disperse the forces.
>
> —*Carl von Clausewitz*[46]

> Cohesion is the art of command.
>
> —*BG (ret.) Eran Dolev*[47]

The development of effective cohesion between soldiers who are expected to fight and work together is probably the most critical ingredient in the prevention of psychological breakdown. Samuel Stouffer

[45] M. T. Hall and M. T. Kennedy, "The Urban Area During Support Missions Case Study: Mogadishu; Applying the Lessons Learned—Take 2," in Glenn (ed.), *Capital Preservation.*

[46] C. Clausewitz, *On War,* M. Howard and P. Paret (eds. and trans.) (Princeton: Princeton University Press, 1984), p. 188.

[47] Dolev interview.

writes in *The American Soldier* that "the group in its informal character served two principal functions in combat motivation: it set and enforced group standards of behavior, and it supported and sustained the individual in stressors he would otherwise not been able to withstand."[48] Col (ret.) Robert Thompson similarly noted that cohesion is part of the "underlying culture. Marines fight for each other. [The] greatest fear to them, greater than dying, is that their buddies will think them as cowards . . . so even if [they're] scared shitless, they just keep going."[49]

Strong unit cohesion is largely a consequence of tough, stressful training. Soldiers learn to push one another and learn that they can rely on the man next to them. Training, especially training that requires group interdependence, builds a bond born of shared experiences. Beyond this, however, commanders should take a personal interest in developing cohesion among their troops. For example, BG Eran Dolev, the surgeon general of the Israeli Defense Forces during the 1982 Lebanon War, notes that commanders must view cohesion as a mission in and of itself and must foster the idea that "we all depend on each other."[50] Moreover, just as marksmanship and physical fitness are part of the indicators of combat readiness, according to Dr. Dolev, commanders must also use morale and cohesion.

Commanders and NCOs should also keep their fingers on the pulse of the unit, identifying problems before they become unmanageable and challenging soldiers to overcome interpersonal squabbles. Soldiers should also be encouraged to spend time with each other outside of organized training events, be it interunit athletic competitions or in on-base housing where units live in close proximity to each other. Development of unit cohesion is especially important for commanders of combat service support units whose MOS often require non-team-oriented work, such as mechanics or truck drivers.

---

[48] Stouffer et al., *The American Soldier*, p. 130.

[49] Col (ret.) Bob Thompson, interview with the author, Fredericksburg, Virginia, February 25, 2003.

[50] Dolev interview.

Special Emphasis on Combat Service Support Units

> Everyone at the staging area is potentially an infantryman: cooks, mechanics, MPs, anyone who is there when bullets fly must be thinking about war. Who would have thought that two Black Hawks would have been shot down in the city? God bless our cooks who stepped up and came to our assistance in the relief convoys. They literally put down the spoons and spatulas and grabbed their weapons to go to the fight. That is the warrior mentality. When we go again, it just might be that low-density military occupational specialty soldier who becomes the last line of defense. He has to be prepared.
>
> —*SFC Matthew Eversmann*[51]

Given that members of combat service support (CSS) units face an increased risk of becoming stress casualties and given the life-threatening risks posed to these units during urban operations, it is critical to see to their training and preparation needs. First, CSS personnel should develop identities as combat-trained soldiers. This in part requires adequate weapons training so that rifles, locked and loaded, can be handled safely and so that marksmanship skills are kept up to date. Live-fire drills will also serve to improve weapon handling and increase confidence. In this latter regard, simunitions are an invaluable tool so that that mistakes will not be life threatening. CSS units should also participate in operational exercises, and situations should be correspondingly developed wherein they actively engage in combat scenarios. They should also train with other CSS and combat units that they will be expected to participate with in combat. "Train as you fight" applies across all MOS designations. Support units should furthermore not be spared the benefits of rigorous PT, especially when strength training can be geared to anticipated duties such as the need for stretcher bearers to climb multifloor stairways as well as move across open fields of fire.

---

[51] Eversmann, "The Urban Area," p. 425.

Postdeployment Battleground Prevention Efforts

Consultation Services

Combat stress control (CSC) units and, more frequently, division mental health assets play an important consultation role. The goal of consultation services is the "enhancement of positive, mission-oriented motivation and the prevention of stress-related casualties."[52] CSC and division mental health personnel can advise unit commanders and other staff such as senior NCOs and chaplains on a variety of prevention-related issues. Consultations may focus on improving unit cohesion, junior leadership skills, or identifying operational stressors or risk factors for battle fatigue. In addition, they can suggest leader actions that would help control stress and help units learn how to best handle cases of battle fatigue. Consultations can be performed across the entire operational time-span from garrison to combat operations to demobilization. Major Michael Doyle notes that the CSC predecessor, called OM teams, were vigorously employed during the Persian Gulf War, where they evaluated unit cohesion and perceived readiness for combat, trained leaders and troops in controlling combat stress, and gave feedback to various command echelons on morale, readiness, and controlling combat stress.[53] Such consultations are a combat multiplier. Commanders and senior NCOs should request their services whenever the need arises.

Rotation and Replacement Policy

World War II and Vietnam taught the military a number of valuable lessons about the consequences of various rotation and replacement policies. In World War II, replacements were introduced on an individual basis and as strangers to those they were supposed to fight with. As a consequence, rates of stress reactions were particularly high

---

[52] Department of Defense, "Chapter 4: Combat Stress Control Consultation." In *Combat Stress Control in a Theater of Operations* (Department of the Army, FM 8-51), http://www.vnh.org/FM851/chapter4.html (last accessed May 22, 2004).

[53] M. E. M. Doyle, "Combat Stress Control Detachment: A Commander's Tool," *Military Review* (May–June 2000), pp. 65–71.

among newly assigned replacement troops.[54] In addition, it was standard policy for soldiers and marines to remain with their units and within their respective theaters for the duration of the war. Because of the war's length and brutality, many men lost hope that they might escape unharmed,[55] and long-term exposure to battle stress resulted in increased rates of stress reactions.[56] In the Vietnam War, as previously discussed, the one-year rotation system created a number of problems. Like World War II, the individual rotation policy impaired unit cohesion, and limiting tours to one year expelled men from units at the height of their combat effectiveness. Problems associated with short-timer's syndrome also arose.

The problems emanating from these two systems and others can be used to derive principles critical to an effective rotation and reconstitution policy. The length of in-theater commitments should be determined by specific criteria, be it the accomplishment of tactical objectives or a set period of time. In either case, soldiers or marines should have a sense that they can survive provided they use their skills wisely. Of course, a rotation policy is completely dependent on the anticipated combat intensity and the number of troops required to accomplish the mission. An out-of-theater rotation policy, for example, may not have been wise during World War II given the manpower shortages of the time.[57]

Intra-theater rotation policies are also critical to maintaining combat effectiveness and minimizing the fallout of psychiatric casualties. Troops constantly exposed to danger wear quickly. Allowing them an opportunity to remove themselves from that danger for more than a few days or weeks enables them to relax, recover their strength,

---

[54] In the 29th Infantry Division, 38 percent of stress casualties were newly arrived replacements. *Headquarters, 29th Infantry Division.*

[55] Statistics from the U.S. Fifth Army bear this out, where only 18 percent of the original members of rifle battalions remained with their units after only 150 days on the Italian front. *Study of AGF Battle Casualties* (Washington, D.C.: Headquarters, Army Ground Forces Plans Section, 1946).

[56] Beebe and Appel, *Variation in Psychological Tolerance.*

[57] R. W. Glenn, *Reading Athena's Dance Card* (Annapolis: Naval Institute Press, 2000).

retrain for the next round of missions, and re-establish military discipline. It is also a time for medals to be awarded to deserving soldiers.[58]

Beyond intra-theater rotation policies, combat commanders, when possible, must be aware of the need to rotate troops in and out of the line of contact. This may be particularly necessary during urban operations that can be physically grueling. According to retired LtGen George R. Christmas, the

> big thing is it depends on the length of time in combat and what I mean by that is . . . urban block-by-block [fighting] lasts 15 to 20 days and by the end of that you'll begin to see who feels the stress. In a company, all platoons are not up in the line at one time, and what is important is continually rotating units . . . rotating units in and out of fight becomes a critical element.[59]

Just as in rotation policies, care should be taken in how replacements are introduced to their units and to combat.[60] If possible, replacement soldiers should not be introduced to their new unit while the unit is engaged in combat operations. Introducing them during the time out of the front lines allows them to train with the unit and develop cohesion prior to entering combat. It is also advantageous if veterans make efforts to train new men in "tricks of the trade" such as how to best prepare MREs (meals ready to eat), use the terrain to their advantage, and understand the behavior of their enemy.[61] Suggestions such as these not only improve soldier survivability but also foster cohesion between new and veteran unit members.

While it is not always feasible to reconstitute depleted units on rotation, the use of replacement teams such as four-soldier units may minimize the stressors of being the odd man out when entering an already cohesive unit. Research during the Korean War, for example,

---

[58] Ibid.

[59] Christmas interview.

[60] Glenn, *Reading Athena's Dance Card.*

[61] Ibid.

demonstrated that entering infantry units in four-man replacement teams resulted in greater levels of morale than when replacements entered one at a time.[62]

Maintaining an Offensive Mindset

> . . . the most dangerous hunt is man on man.
>
> —*LtCol (ret.) John Allison*[63]

It is critical for soldiers and marines, whether in urban environments or not, to maintain a sense of situational awareness. In urban-based peacekeeping operations, soldiers can quickly develop a mindset that they are merely passive targets to could-be lone gunmen behind a distant doorway or window frame. Such circumstances can easily engender a sense of helplessness that would clearly lead to increased stress levels and possible breakdown. The answer, according to John Allison, is to maintain the mindset of the hunter and not the hunted. He noted that marines in Mogadishu patrolled with an aggressive posture. In addition to enforcing the ban on carrying illegal arms, "we would run sweeps from the Bacara arms market [and] we never let them get their heads up." Allison goes on to state:

> that in and of itself helped. Probably why we didn't see [stress reactions] in the Marine Corps. They got in the rhythm and . . . did things by virtue of their training, and they didn't feel like targets. . . . So if you were out there and kept moving things, stayed engaged and kept your mind active, it gave a sense that you were in control. More like you were the hunter and they were the hunted.[64]

---

[62] D. J. Chester and N. J. Von Steenberg, "Sociometry in the Armed Forces: Effect on Morale of Infantry Team Replacement and Individual Replacement Systems," *Infantry Team Replacement,* Vol. 18, No. 4 (1955), pp. 331–341.

[63] Allison interview.

[64] Ibid.

Just as marines in the Battle of Hue felt an enhanced sense of control in their ability to directly target enemy forces, maintaining an aggressive and proactive posture in stability operations will protect soldiers and marines alike from a lost sense of control.

Intelligence

> Everyone said afterwards "they knew that part of town was bad." Thanks for telling me.
>
> —*SFC Matthew Eversmann referring to Task Force Ranger raid, October 3, 1993, Mogadishu, Somalia*[65]

Intelligence—knowing what to expect—plays an important role in stress mitigation. According to WO2 Kevin Lawrence, "nothing's worse than going into the unknown; going into the unknown can raise the stress level. Concern about the unknown also distracts the mind's attention from the mission at hand to a vast array of potential and often times worst-case scenarios, if someone doesn't get the whole picture it leaves room to create their own."[66] A lack of intelligence also raises many tactical problems that can lead to stress such as an ambush that could have been anticipated or, according to Robert Barrow, engagements precipitated by cultural misunderstandings. Failure to understand that many Somalis view their sidewalk as an extension of their home resulted in several confrontations during Operation Restore Hope that could have been avoided.[67] Intelligence is thus key to reducing the stress and distraction inherent in combat or stability operations. It has been demonstrated as early as World War II that combat aircrews experienced fewer stress reactions when informed in advance about incoming enemy air attacks.[68] Psychiatrist

---

[65] Eversmann, "The Urban Area," p. 420.

[66] WO2 Kevin Lawrence, interview with the author, Washington, D.C., July 1, 2003.

[67] Barrow interview.

[68] I. L. Janis, *Air War and Emotional Stress* (New York: McGraw-Hill, 1951), referenced in Driskell and Johnston, "Stress Exposure Training."

and LtCdr Neil Greenberg of the Royal Navy argues that "the more you know about what you're going to encounter, what the sounds will be like, the weaponry you expect to have, all those things will work to reduce stress."[69] From a commander's perspective, LtCol (ret.) John Allison states that "The more we can share with the guy at the end of the spear, the better he can perform the mission he's been given and the better the chances are of him coming back alive."[70]

A significant amount of information must be disseminated to troops before entering a combat zone. Prior to entering a theater of war, soldiers or marines must have some understanding of the indigenous culture, and they must be familiarized with the types of weapons used by the enemy and the sounds of those weapons. They should also learn enemy tactics, and, importantly, they should have an understanding of why they are being sent to fight in the first place. Mission- or patrol-specific information should include mission objective, threats known or suspected, and the location of sympathizers. Critically, according to WO2 Kevin Lawrence, soldiers and marines should always have a "plan B [or an] extraction plan, [it] gives the soldier a comforting feeling that if all else fails this is what we do."[71]

Rules of Engagement

> This insanity, these damnable rules of engagement that prevented American fighting men from using the only tactical assets that gave us an advantage during firefights—that of our vastly superior fire power represented by air strikes, artillery, and naval gunfire—these orders continued to remain in force and hinder, wound, and kill 1/5 Marines until the fourth day of fighting inside the citadel of Hue.
>
> —*Nicholas Warr*[72]

---

[69] Greenberg interview.

[70] Allison interview.

[71] Lawrence interview.

[72] Warr, *Phase Line Green*, p. 135.

> From a morale perspective, soldiers who . . . have the decision and know they can shoot back do better than soldiers that are in peacekeeping [operations] standing there while Shiites rain mortars down on them.
>
> —*MAJ Michael Doyle*[73]

Complex and overburdensome rules of engagement not only threaten the physical safety of U.S. military personnel but may also engender a sense of vulnerability, helplessness, and a lack of confidence in oneself and in leadership, problems that may presage acute or long-term psychological problems.

Soldiers and marines should have all the resources they need to successfully prosecute war and defend against threats to their safety. For combat operations, U.S. servicemen should have the full complement of supporting arms at their disposal. While the direct targeting of civilian populations must be limited, it should not necessarily be ruled out, particularly if such restrictions endanger American lives and jeopardize the overall campaign. For peacekeeping operations or checkpoint operations, ROE should be designed to be as clear-cut and simple as possible. To lessen reaction times in the event deadly force is necessitated, ammunition rounds may have to be chambered. Notes John Allison, "if you know you don't have the round chambered, that's one more thing you have to do. If you have the round chambered your confidence is that much higher. You're ready and . . . word [of the chambered round] gets out on the street." Obviously, adoption of such a policy would have to be preceded by extensive training and accompanied by constant refreshers to ensure maintenance of discipline and the safety of soldiers and local civilians.

Prior to operations, varying scenarios related to ROE can be practiced in tactical exercises, allowing leaders and soldiers alike to practice decisionmaking skills. COL Robert Collyer noted that a similar program worked well for Australian soldiers preparing for stability operations in East Timor and Rwanda and suggested that such

---

[73] Doyle interview, reference to peacekeeping operations in Beirut, Lebanon.

training also gives soldiers and leaders confidence in their ability to do what is required of them.[74]

Physiological Factors

Sleep. Sleep deprivation is an inherent risk during combat operations. In addition to increasing soldier risk for battle fatigue, sleep deprivation impairs performance in a number of domains. Every 24 hours of sleep loss impairs mental performance by 25 percent, most notably in domains related to complex mental processes such as situational awareness (knowing where you are on the battlefield), information integration (e.g., integrating disparate battlefield reports into a single mosaic), and the quick processing of information (e.g., translating a fire mission to the actual firing of an artillery round).[75] Sleep loss also increases the body's natural drive for sleep, which in turn impairs vigilance[76] (e.g., failure to detect approaching threats) and increases the likelihood of unintended sleep intrusions[77] (e.g., nodding off while on watch). Impairments in these domains can be seen over time following only brief deficits in sleep time, such as several hours a day over a period of a week or two. Dramatic impairments can be found after 24 hours of sleep loss.[78] Importantly, commanders are especially vulnerable to such impairment. Studies have shown that operationally based sleep loss increases as a function of chain of command.[79]

---

[74] Collyer interview.

[75] G. L. Belenky, *Sleep, Sleep Deprivation, and Human Performance in Continuous Operations*, http://www.usafa.af.mil/jscope/JSCOPE97/Belenky97/Belenky97.htm (last accessed November 21, 2003).

[76] G. Belenky, N. J. Wesensten, et al., "Patterns of Performance Degradation and Restoration During Sleep Restriction and Subsequent Recovery: A Sleep Dose-Response Study," *Journal of Sleep Research,* Vol. 12, No. 1 (2003), pp. 1–12.

[77] M. Harma, S. Suvanto, et al., "A Dose-Response Study of Total Sleep Time and the Ability to Maintain Wakefulness," *Journal of Sleep Research,* Vol. 7, No. 3 (1998), pp. 167–174.

[78] L. Rosenthal, T. A. Roehrs, et al., "Level of Sleepiness and Total Sleep Time Following Various Time in Bed Conditions," *Sleep,* Vol. 16, No. 3 (1993), pp. 226–232.

[79] As reviewed in Belenky, *Sleep, Sleep Deprivation, and Human Performance.*

Commanders should insist on proper operational sleep routines for both themselves and those in their charge. Clearly, seven to eight hours of sleep per 24-hour cycle is ideal. When operational requirements do not permit this, four or more uninterrupted hours of sleep per 24-hour cycle may be acceptable, but only for one or two weeks, after which the sleep debt should be repaid.[80] Even sleep obtained in brief naps (20 minutes to 2 hours) can improve alertness and performance,[81] though a brief period of post-nap grogginess should be anticipated for naps beyond 20 minutes. Sleep for longer durations should take place at regular scheduled times during the 24-hour day in order to maintain a stabilized circadian rhythm. Soldiers must understand that ambient noise such as shellfire or the conversations in a command tent will make sleep more fragmented and less effective.[82] When possible, soldiers and marines should sleep in a quiet place. Finally, combat operations and the rotation of troops to and from the line of contact should be scheduled according to set shifts that permit soldiers time to sleep and eat when possible. Twelve-hour on and off schedules are ideal, though six- and four-hour shifts can work as well if properly designed.[83]

Stimulants such as caffeine[84] and amphetamines[85] can temporarily reverse sleep-induced impairments in sleepiness and performance. While amphetamine pills have been used to help pilots maintain alertness during long, monotonous flights, the ground soldier typically uses caffeinated coffee or recently developed caffeinated

---

[80] Department of Defense, *Leaders' Manual for Combat Stress Control.*

[81] T. Helmus, L. Rosenthal, et al., "The Alerting Effects of Short and Long Naps in Narcoleptic, Sleep Deprived, and Alert Individuals," *Sleep,* Vol. 20, No. 4 (1997), pp. 251–257.

[82] M. H. Bonnet, "Sleep Restoration as a Function of Periodic Awakening, Movement, or Electroencephalographic Change," *Sleep,* Vol. 10 (1987), pp. 364–373.

[83] Department of Defense, *Combat Stress.*

[84] L. Rosenthal, T. Roehrs, et al., "Alerting Effects of Caffeine after Normal and Restricted Sleep," *Neuropsychopharmacology,* Vol. 4, No. 2 (1991), pp. 103–108.

[85] C. Bishop, T. Roehrs, et al., "Alerting Effects of Methylphenidate Under Basal and Sleep-Deprived Conditions," *Experimental and Clinical Psychopharmacology,* Vol. 5, No. 4 (1997), pp. 344–352.

formulations of chewing gum or candy bars. It must be understood that stimulants do not replace or reduce the body's natural drive for sleep, they only mask it. Following periods of sleep deprivation, the body will eventually require sleep, and the longer the sleep deprivation, the longer period of necessary replenishment sleep.[86] It is important to note that stimulants increase the risk of anxiety reactions, as evidenced by laboratory and clinical experiments.[87] Thus, while caffeine may improve some performance measures in sleep-deprived individuals, it may have a corresponding effect of increasing the risk of acute stress reactions. The best approach to maintaining an alert and high-functioning combat team is for command to ensure that they and their men obtain an adequate amount of sleep.

Nutrition, Water, and Load. Combat on any terrain consumes significant amounts of calories and increases the likelihood of dehydration. Soldiers and their leaders must pay careful attention to their caloric intake and hydration practices to ensure their utmost combat effectiveness and protection against battle fatigue. Given sufficient supplies, "food and water discipline" may take on the meaning of eating and drinking enough rather than limiting intake to preserve scarce resources. One soldier from TFR recalled "no one told me about drinking water . . . When you do get in that big firefight you want to make damn sure that they're drinking water and make sure they're eating too."[88] Unfortunately, even with proper food intake discipline, the calories available in an MRE diet may fall short of what a soldier needs.[89] Research should be conducted to accurately determine caloric demands during urban operations, with meal contents designed accordingly. Research is presently being conducted on novel nutrient delivery systems such as the Transdermal Nutrient

---

[86] Rosenthal et al., "Level of Sleepiness."

[87] I. Iancu, O. T. Dolberg, and J. Zohar., "Is Caffeine Involved in the Pathogenesis of Combat-Stress Reaction?" *Military Medicine,* Vol. 161, No. 4 (1996), pp. 230–232.

[88] MSgt Matthew Eversmann, interview with the author, Baltimore, Maryland, May 22, 2003.

[89] Glenn, Hartman, and Gerwehr, *Urban Combat Service.*

Delivery System, which uses a skin patch to transfer nutrients to a soldier via a "microdialysis" process.[90]

Excessive loads can also deplete the physical energy necessary to conduct combat operations. According to COL Gregory Belenky, load is the most important physiological factor, more important than hydration, sleep, or nutrition.[91]

Maintaining Morale

Soldiers and marines with high levels of morale are protected against stress-induced breakdown. Building strong battleground morale is not exceptionally difficult, as many of the contributing factors are elementary. The factors that contribute to personal morale were studied during the first Gulf War. According to one of the authors, "in this study, soldiers identified mail, showers, tents, rest areas, hot food, cold drinks, being able to live as squads, crews, or platoons in self-improved areas, entertainment, and free time as being significant contributors to morale. Also contributing were visits to the resort area at Half Moon Bay and phone calls home."[92] Mail is an especially important element. "You need to do whatever you can to get mail to those folks. Those with girlfriends and parents . . . that is such a motivating thing for those warriors to fight knowing there is something back there. That is really important."[93]

While many of the factors identified above are the responsibility of command, the role of the small unit leader in maintaining morale cannot be overstated. In Operation Iraqi Freedom, an unidentified Marine first sergeant submitted an informal lessons-learned list for NCOs while en route from the Gulf. What follows are some of his suggestions that pertain to unit morale and illustrate the independent power of NCOs to sustain their men in combat:

---

[90] Ibid., p. 91.

[91] Belenky interview.

[92] Wright, *Operation Desert Shield Desert Storm*, p. 22.

[93] Lawrence interview.

> Buy a short wave radio and get the news. Write it down under a poncho at 0200. Get the baseball scores out to the Marines and you are a hero . . . .
>
> Have all the e-mail addresses of your Marines' wives. Get to any HHQ and send a blanket e-mail to all of them . . . .
>
> Ensure your Marines write letters on anything they can get their hands on. MRE boxes work great. I put a ammo can on my vehicle for outgoing mail. Get the mail out. There is always a way. Pass if off to other units if you have to. Find a helo and give him your mail. Give him a can of dip to do it for you . . . .
>
> Use the SAT Phone. Forget the cost. Grab a few young Marines when you can and let them call home. That Marine could lead the entire battalion after he talks to his wife after a firefight . . . .
>
> Field Hygiene. Marines got sick. Some pretty bad. Look at your Marines daily if you can. Ask questions. Marines will not tell you they are sick until they go down hard. They are a proud bunch.[94]

Another element of maintaining morale is to help soldiers develop a sense of closure with their fallen comrades. In this regard, battlefield memorials are key. LtGen (ret.) Christmas described his experience with a memorial service following the loss of several men in a firefight.

> The company was really down because first of all they had lost these young marines. Actually in their subconscious they felt they had run away in the face of the enemy because we were forced back across the ridge. And there was psychological damage there. Fortunately, there was a good chaplain, a Catholic priest, and they put us in reserve in the Christian Brothers school . . . and he found this chapel and he offered a memorial mass, and what occurred there was a closure with those we had

---

[94] R. W. Glenn, *AAR From the Warlords [24 Marine Expeditionary Unit]* (May 7, 2003).

> just lost and . . . so those kinds of things happen continually in combat, and that is clearly a stress-related type aspect.[95]

Post-Operational Recommendations

Debriefing

S.L.A. Marshall was the first practitioner and advocate of after action debriefing. In this formulation of debriefing, soldiers gather in small unit based groups and discuss their operational experiences and subsequent reactions. Through this interaction, soldiers come to realize that each individual experiences the same event in very different ways. The goal of this approach is not to limit post-combat psychiatric reactions such as PTSD, but to develop a "historical truth" about what occurred during the course of a mission and thus re-establish group unity.[96] It also fosters a discussion of small unit lessons learned that may improve subsequent combat effectiveness. On the former point, the psychiatrist Gregory Belenky comments:

> You get with debriefing a common coherent picture, rather than thinking nine people let you down, you realize it's only one. An individual person can think several things went wrong [and through debriefing] realize that these things are incompatible. You get a common coherent narrative.[97]

In contrast to Marshall's approach, psychological debriefing (PD) is a clinical tool that seeks to reduce psychological distress, such as post-traumatic stress disorder (PTSD), that may follow exposure to traumatic events. The treatment is conducted militarily through a single group session in which participants discuss their individual experiences of a shared traumatic event and describe their emotional

---

[95] Christmas interview.

[96] Z. Kaplan, I. Iancu, and E. Bodner, "A Review of Psychological Debriefing after Extreme Stress," *Psychiatric Services*, Vol. 52, No. 6 (2001), pp. 824–827.

[97] Belenky interview.

reactions. The facilitator, usually a person from outside the group who has received specialized PD training, then attempts to normalize the group's feelings as something that is universal and expected, and future emotional reactions are also discussed. The U.S. military regularly administers psychological debriefings to its servicemen following traumatic exposures. The most commonly used form of PD is called Critical Incident Stress Debriefing (CISD) and is generally used by the U.S. Navy and Marine Corps. The U.S. Army uses a modified version of CISD, which they call Critical Event Debriefing (CED).[98]

Unfortunately, scientific evidence corroborating the effectiveness of PD is controversial at best. In randomized controlled trials (RCTs), individuals for whom the intervention is designed to help (e.g., those exposed to a traumatic event) are randomly and blindly assigned to a condition containing an active treatment (e.g., PD) or to a condition containing an inert or inactive condition (e.g., simple education, usually referred to as a placebo). In the event of PD, if PD compared with education results in lower rates of PTSD, say 6–12 months following treatment exposure, then investigators may conclude that, at least in this instance, PD as a preventative measure for PTSD is effective. [99] If multiple RCTs, conducted by multiple investigators, are shown to produce similar outcomes, then one can broadly argue that PD works as it is intended to work.

But by many accounts, PD does not meet this standard. Several scientific reviews of PD RCTs (including CISD or CISD-like treatments) suggest that PD does not significantly reduce the likelihood of PTSD following exposure to traumatic events. One such review tabulated the results of nine studies that used randomized designs and incorporated single-session PD treatments. The authors of this review concluded "there is no current evidence that psychological debriefing

[98] Stokes, written comments. CED follows the CISD format, though greater flexibility in its administration is allowed and some approximations are made toward Marshall's historical group debriefing.

[99] The use of RCTs as described is especially important for debriefing because open trials—those that do not rely on control groups but simply monitor the progress of a single treated group—will show efficacy because most trauma-exposed individuals improve over time. Brett Litz, Written comments, December 18, 2004.

is a useful treatment for the prevention of post-traumatic stress disorder after traumatic incidents. Compulsory debriefing of victims of trauma should cease."[100] In another review, van Emmerik and colleagues conducted a meta-analysis[101] on seven studies (five of which were RCTs) of CISD and non-CISD interventions. Summarizing their findings, the authors state: "Thus, CISD was no more effective than non-CISD interventions or even than not intervening at all."[102] Finally, in another review, McNally and colleagues conclude that "RCTs of individualized debriefing and comparative, nonrandomized studies of group debriefing have failed to confirm the method's efficacy. . . . For scientific and ethical reasons, professionals should cease compulsory debriefing of trauma-exposed people."[103]

Beyond the mere finding that PD does not work, some studies also suggest that PD might also be harmful. In the van Emmerik study noted above, effect sizes, a statistical measure of the robustness of a given outcome, were even smaller (and thus less robust) for CISD interventions than non-CISD interventions and no-intervention controls.[104] In an additional review, two well-conducted studies of six RCT studies provided evidence of greater PTSD symptoms at follow-up for PD versus control groups. Despite this finding, the authors of this review caution that "it is premature to conclude unequivocally that PD hinders recovery from trauma."[105]

---

[100] S. Rose, J. Bisson, and S. Wessely, "Psychological Debriefing for Preventing Post-Traumatic Stress Disorder (PTSD) (Cochrane Review)," p. 1, in *The Cochrane Library, Issue 4, 2003* (Chichester, UK: John Wiley & Sons, Ltd.).

[101] Meta-analysis is a statistical method in which outcomes from varying studies are combined and evaluated statistically.

[102] A. A. van Emmerik, J. H. Kamphuis, et al., "Single Session Debriefing After Psychological Trauma: A Meta-Analysis," *Lancet,* Vol. 360, No. 9335 (2002), p. 769.

[103] R. J. McNally, A. B. Richard, and A. Ehlers, "Does Early Psychological Intervention Promote Recovery from Posttraumatic Stress?" *Psychological Science in the Public Interest,* Vol. 4, No. 2 (2003), p. 72.

[104] van Emmerik et al., "Single Session Debriefing."

[105] B. T. Litz, M. J. Gray, R. A. Bryan, A. B. Adler, "Early Intervention for Trauma: Current Status and Future Directions," *Clinical Psychology: Science and Practice,* Vol. 9, No. 2 (2002), p. 124.

Many advocates of PD, particularly CISD advocates, argue that the studies reviewed above are not representative of CISD in its intended format: for example, therapists were not properly trained in the CISD approach, or the treatment was not delivered according to CISD guidelines and the intervention was often administered to individuals or groups who lacked a common experience (and thus would not be representative of its use in military populations).

These criticisms call for continued PD research, particularly in military settings and with the PD platforms commonly used by the military.[106] Nonetheless, it still stands that the efficacy of PD remains unproven. Until a substantial amount of new research supporting it comes along, the best practice for the U.S. military may be to follow the science and cease psychological debriefing altogether.[107]

Summary

> It's far more important to get this in major modules in the officer and senior NCO educational systems. Field manuals are lovely, but many people never read them. And for people to retain [the information], they've got to know that someone gives this a reasonably high priority. And if you've got a module at officer and NCO courses, there is a greater recognition that this is important. And in the Army, the other responsibility is continuing education as part of the division mental health team . . . . I think it's a critical recommendation . . . . It should be placed at every

[106] This would include the need to conduct RCTs on the Army CISD variant, CED. To the extent that this is a uniquely different PD platform, it is noteworthy that its effectiveness has never been evaluated in a randomized controlled trial. COL Elspeth C. Ritchie, M.D., written comments, December 14, 2004.

[107] Not all forms of intervention lack merit. Two promising approaches include multisession cognitive behavioral therapy for trauma survivors and an intervention program developed by Captain Cameron March of the British Navy. CBT has demonstrated effectiveness in RCTs, though studies with military populations remain to be conducted. S. Wessely, "Psychological Debriefing Is a Waste of Time," *British Journal of Psychiatry,* Vol. 183 (2003), pp. 12–14. Litz, written comments.

> level of the education system to be constantly reinforced like basic on up, like the initial NCO courses on up. The question is making it part of the tool kit, if you will, of every leader: some knowledge as to what it looks like, what it does to people, what the realities are, and what we can do to prevent it and treat people.
>
> —*David H. Marlowe, Ph.D.*[108]

Although CSR is not 100 percent preventable, there are a number of steps available to commanders that could reduce the incidence of negative combat stress reactions. Critically, as suggested throughout this chapter, actions that reduce CSR have a correspondingly positive impact on combat effectiveness. The two outcomes are inextricably linked. From a garrison perspective, new soldiers should be given a proper introduction to their new unit's history and lore. Officers and NCOs should make a cohesive fighting force a top priority. Importantly, training should be rigorous and should expose soldiers to the stressors they are likely to face in combat and during other military operations while at the same time allowing them to practice anticipated skills. Reality-based operational exercises are key in this regard. Combat support and, notably, combat service support units require such training no less than combat units, given their vulnerability and inevitable exposure to combat.

While a soldier goes to war with the skills he or she learned while in garrison, there are a number of critical factors that must be addressed that can offset the risk of battle fatigue. Division mental health teams can provide an immeasurable service to commanders by evaluating the unit's social climate and morale and suggesting specific actions available to commanders that may limit CSR risk. While rotation and replacement policies are obviously critical to unit cohesion and fighting effectiveness, commanders should not underestimate the value of communicating relevant intelligence findings to their sol-

---

[108] Marlowe interview.

diers. Commanders should also understand that offensive operations give soldiers confidence and a sense of control over their environment. While an obvious statement in regard to combat operations, an offensive mindset can also have benefit during stability operations. ROE should be simple and designed so that soldiers and marines have the means necessary to defend themselves. Finally, commanders and NCOs should ensure that their men stick to sleep schedules and eat and drink appropriately throughout operations. While all these factors inevitably contribute to morale, other factors such as hot food, showers, mail, and access to leisure activities communicate to the soldier that command is looking after him.

The authors suggest that combat stress control measures be taught as separate modules in officer and NCO training programs. Such training would lead to a greater appreciation of the prevalence and risk of battle fatigue and should work to improve its prevention and treatment.

The suggestions summarized above can be applied to nearly any operational terrain; however, some may have particular relevance to urban operations. For example, the stress exposure training program that culminates in tough and realistic combat exercises can be especially tailored for soldiers preparing for urban operations. Once committed to the urban operation, a rotation policy that gives soldiers a needed rest from the physical and emotional burdens of highly intense combat operations is important, as is the need to maintain nutrition and water intake during the physically strenuous operations. Suggestions regarding intelligence, rules of engagement, and offensive operations also have particular relevance toward urban operations, especially those that are stability focused.

CHAPTER SEVEN

Conclusions and Recommendations

The manpower loss imposed by combat stress casualties creates a significant burden to combat and CSS units. In addition, it is certain that in the future the U.S. military will be committed to operations in urban environments. Such undertakings will encompass all the stressors found in other operating environments, often more intensely, and other stressors rarely confronted in less-populated terrain. The authors of this report have sought to evaluate the treatment and prevention of stress casualties and their implications for urban operations. This final chapter summarizes the study's primary findings and reviews recommended actions.

Risk of Stress Casualties in Urban Operations

Soldiers and marines with operational experience in the Battle of Hue, Operation Just Cause, and Somalia stability operations testify to the fact that urban operations are highly stressful endeavors. Likewise, many medical and scholarly authorities suggest that such stressors would translate into increased rates of combat stress reaction. Reasons for these assertions seem obvious. In addition to the many tactical stressors inherent in a three-dimensional battleground with many fields of fire, close-quarters fighting, and restrictive rules of engagement, the urban environment also intensifies many of the environmental factors known to engender stress casualties. Such factors include high casualty rates, inordinate dispersal of forces that strain both unit cohesion and trust in leadership, and the necessity that CSS units operate close to and often in the midst of combat operations.

For the above reasons, rates of stress casualties in prior urban operations were expected to be high. However, the findings from our review of the World War II battles of Brest and Manila, Vietnam's Battle of Hue, and other urban conflicts do not bear this out. In each of these battles, available evidence indicates that rates of combat stress reactions were not higher than those incurred during operations on other types of terrain.[1]

There thus appears to be a general discrepancy between reported levels of subjective stress and documented rates of stress casualties. To account for this, it is suggested that urban operations instill a degree of control in combatants who have the power to individually engage enemy personnel. The mental strain involved in close-quarters fighting may also serve to distract soldiers from subjective levels of stress that are recalled in hindsight. The nature of offensive operations may have also limited the frequency of stress reactions. Finally, we observed that a hostile civilian presence was lacking in each of these operations, as were enemy combatants who disguised themselves in civilian clothing. Many medical authorities have strongly suggested that civilian casualties may be a major precipitating factor for short- and long-term stress reactions. As such, with the exception of the relatively brief Battle of Jenin, our data do not speak to combat operations waged by conventional forces in operational environments with high and potentially combative civilian populations.

Treatment of Combat Stress Reactions

Combat stress reaction is referred to on the battlefield as battle fatigue, a term that emphasizes the need for rest as the primary restorative treatment. Briefly, the symptoms of battle fatigue are varied, they can form multiple symptom constellations, and those constellations

[1] The only exception to this may be CSR rates for the Battle of Jenin, which were relatively high given the battle's brief duration.

can change dramatically over time. Best practice is to beware of persistent and progressive decline in soldier mood and performance. Treatment is characterized by the acronym PIES: proximity (treat soldier close to the front), immediacy (treat soon after symptom onset), expectancy (reassure that soldier is not ill and that he will return to duty), and simplicity (ensure that soldier has food and drink, and bring body temperature to a normalized state). While these terms illustrate treatment by combat stress control units, they also apply to treatment within the operating unit (provided the tactical situation permits). While most soldiers with a prior episode of CSR do well once symptoms recede, others will go on to have a second reaction. The outlined treatment principles apply to operations on urbanized terrain and, in fact, may be simplified by the close proximity of medical support units to the lines of contact. Their applicability becomes limited, however, during fast-moving operations or operations involving deep spaces of battle. As currently configured, combat stress control units are also hindered because they are unknown by the line communities they serve. It is consequently suggested that the U.S. military adopt peer mentoring programs that provide NCOs training in stress control as organic to maneuver units. Organically assigning mental health staff to maneuver brigades, a process already initiated in the U.S. Army, should continue.

Recommended Prevention Efforts

Many, though not all, stress reactions can be prevented before they start. This report described many preventive measures available to commanders and NCOs. These suggested actions include the following:

- **Soldier indoctrination.** Commanders and NCOs should attempt to build a degree of unit loyalty and pride in new members through teaching new members the unit history and lore. Liberal use of symbolic and historical artifacts can further enhance this lesson.

- **Tough and realistic training.** This report advocates a model of stress exposure training developed by James Driskell and Joan Johnston. These authors suggest a three-tiered level of training. First, soldiers and marines should be taught about the variety of stressors inherent in operational environments, how they might respond to those stressors, and how it may impact their performance. Skill sets necessary for anticipated operation should then be taught. Specific training approaches reviewed include helping soldiers limit distracting and non-goal-relevant thoughts, practicing skills amidst stressful conditions, pairing tasks that are expected to co-occur, over-learning tasks through repetition, and practicing team skills under duress. Finally, to improve soldier confidence, learned tasks should be practiced under conditions of increasing stressful stimuli that approximate those likely to be encountered operationally.
- **Cohesion.** This report suggests that cohesion ought to be a priority for commanders and NCOs. Specific ways to improve cohesion include stressful training that demands teamwork from soldiers who are expected to work closely together and increasing non-duty-oriented interactions such as athletic programs, social events, and unit centered housing. Identifying and remedying interpersonal problems before they worsen is also helpful.
- **CSS-specific suggestions.** Combat service support units should enhance combat-specific skills and develop identities as combat soldiers. Stress exposure training suggestions identified above are also applicable to CSS units. In addition, building unit cohesion should be a special focus for CSS commanders.
- **Consultation services.** Combat stress control units and division mental health teams can help commanders identify and remedy problems with unit-based morale and cohesion and provide seminars on stress and stress reactions. These units can also consult with commanders on unit reconstitution, on operational risk factors for stress reactions, and on available leader actions to offset those risk factors. Commanders should utilize their services, especially prior to and during operations.

- **Rotation and reconstitution.** Out-of-theater rotation policies should be communicated as early as possible to soldiers. Time in theater should not be so long as to limit a soldier's confidence in survivability. Intra-theater rotations should be employed such that entire units are withdrawn from the line of contact for at least several weeks and permitted an opportunity to relax, retrain, and reassert military decorum. Units should also be rotated in and out of the line of contact during urban operations. Replacement policies are best when cohesive groups of soldiers enter into a unit withdrawn from battle. Policies such as these enhance unit cohesion and limit unremitting exposure to combat.
- **Offensive operations.** During military operations other than war (MOOTW), it is suggested that units maintain an aggressive and proactive posture. In addition to its tactical benefits, such operations will serve to increase soldiers' sense of control over their environment.
- **Intelligence.** Keeping soldiers informed reduces stress. Soldiers should be briefed on relevant indigenous cultures and enemy weapons and tactics. Intelligence relevant to individual missions includes mission objective, threats known or suspected, location of sympathizers, and a specific exit plan.
- **Rules of engagement.** For stability operations, ROE should be clear-cut and simple. Rounds may have to be chambered to lessen reaction times to danger. Strong and ethical leadership, allowing soldiers the opportunity to practice ROE in training exercises, and proper weapons-handling training, both increases soldier confidence in their ability to interpret ROE and limits the extent of ROE violations.
- **Physiological factors.** Sleep is important for both maintaining alertness and cognitive performance and minimizing the risk of stress casualties. As such, commanders should ensure that they and their soldiers practice good sleep hygiene. Eight hours of sleep per day is best, though four hours of uninterrupted sleep

for short stretches of time may do. Noisy environments impair sleep and thus increase the need for longer sleep durations. Maintaining proper fluid and nutrition intake is likewise important. Soldiers should drink fluids throughout their operations. MREs may fall short of required caloric intake. Research should be conducted on the caloric demands of urban operations.

- **Morale.** Maintaining unit morale is the job of both commander and NCO. Efforts should be made to ensure that soldiers have access to morale-boosters such as mail, showers, hot food, cold drinks, occasional calls home, and entertainment. Timely memorials for the fallen are also key.
- **Debriefing.** After action debriefings that review operational lessons learned and help soldiers develop an understanding of the events of combat operations may be helpful. In contrast, psychological debriefings that attempt to prevent chronic stress reactions such as PTSD have not been proven effective.
- **Education.** Information on the nature, prevention, and treatment of combat stress reactions should be introduced as modules in officer and NCO training programs.

Bibliography

Journal Articles

Artiss, K. L. "Human Behaviour Under Stress: From Combat to Social Psychiatry," *Military Medicine,* Vol. 128 (1963), pp. 1011–1015.

Bacon, B. L. and J. J. Staudenmeijer. "A Historical Overview of Combat Stress Control Units of the U.S. Army." *Military Medicine,* Vol. 168, No. 9 (2003), pp. 689–693.

Bandura, A. "Self-Efficacy: Toward a Unifying Theory of Behavioral Change." *Psychological Review,* Vol. 84, No. 2 (1977), pp. 191–215.

Belenky, G., N. J. Wesensten, D. R. Thorne, M. L. Thomas, H. C. Sing, et al. "Patterns of Performance Degradation and Restoration During Sleep Restriction and Subsequent Recovery: A Sleep Dose-Response Study." *Journal of Sleep Research,* Vol. 12, No. 1 (2003), pp. 1–12.

Bishop, C., T. Roehrs, L. Rosenthal, and T. Roth. "Alerting Effects of Methylphenidate Under Basal and Sleep-Deprived Conditions." *Experimental and Clinical Psychopharmacology,* Vol. 5, No. 4 (1997), pp. 344–352.

Block, H. S. "Army Clinical Psychiatry in the Combat Zone: 1967–1968," *American Journal of Psychiatry,* Vol. 126, No. 3 (1969), pp. 289–298.

Blood, C. G., and M. E. Anderson. "The Battle for Hue: Casualty and Disease Rates During Urban Warfare." *Military Medicine,* Vol. 159, No. 9 (1994), pp. 590–595.

Blood, C. G., and E. D. Gauker. "The Relationship Between Battle Intensity and Disease Rates Among Marine Corps Infantry Units." *Military Medicine,* Vol. 158, No. 5 (1993), pp. 340–344.

Bonnet, M. H. "Sleep Restoration as a Function of Periodic Awakening, Movement, or Electroencephalographic Change." *Sleep,* Vol. 10, No. 4 (1987), pp. 364–373.

Bourne, P. G. "Military Psychiatry and the Vietnam Experience." *American Journal of Psychiatry,* Vol. 127, No. 4 (1970), pp. 481–488.

Brill, N. Q., G. W. Beebe, and R. L. Lowenstein. "Age and Resistance to Military Stress." *U.S. Armed Forces Medical Journal,* Vol. 4, No. 9 (1953), pp. 1247–1266.

Chester, D. J., and N. J. Von Steenberg. "Sociometry in the Armed Forces: Effect on Morale of Infantry Team Replacement and Individual Replacement Systems." *Infantry Team Replacement,* Vol. 18, No. 4 (1955), pp. 331–341.

Chupick, D. M. L. "Training for Urban Operations." *Dispatches: Lessons Learned for Soldiers,* Vol. 9, No. 2 (2002), pp. 3–42.

Dekel, R., Z. Solomon, K. Ginzburg, and Y. Neria. "Combat Exposure, Wartime Performance, and Long-Term Adjustment Among Combatants." *Military Psychology,* Vol. 15 (2003), pp. 117–131.

Doyle, M. E. M. "Combat Stress Control Detachment: A Commander's Tool." *Military Review,* May–June 2000, pp. 65–71.

Driskell, J. E., J. H. Johnston, and E. Salas. "Does Stress Training Generalize to Novel Settings?" *Human Factors,* Vol. 43 (2001), pp. 99–110.

Driskell, J. E., R. P. Willis, and C. Copper. "Effect of Overlearning on Retention." *Journal of Applied Psychology,* Vol. 77 (1992), pp. 615–622.

Egan, J. R., L. Jackson, and R. H. Eanes. "A Study of Neuropsychiatric Rejectees." *Journal of the American Medical Association,* Vol. 145 (1951), pp. 466–469.

Fontenot, G. "Fear God and Dreadnought: Preparing a Unit for Confronting Fear." *Military Review,* July–August 1995, pp. 13–24.

French, C. E., R. J. Rona, M. M. Jones, and S. Wessely. "Screening for Physical and Psychological Illness in the British Armed Forces: II. Barriers to Screening—Learning from the Opinions of Service Personnel." *Journal of Medical Screening,* Vol. 11, No. 3 (2004), pp. 153–161.

Gabriel, R. F., and A. A. Burrows. "Improving Time-Sharing Performance of Pilots Through Training." *Human Factors,* Vol. 10 (1968), pp. 33–40.

Gal, R., and F. D. Jones. "A Psychological Model of Combat Stress." In F. D. Jones, L. R. Sparacino, V. L. Wilcox, J. M. Rothberg, and J. W. Stokes (eds.), *War Psychiatry* (Washington, D.C.: TMM Publications, 1995), pp. 133–148.

Glass, A. J. "Effectiveness of Forward Neuropsychiatric Treatment." *Bulletin of the U.S. Army Medical Department,* Vol. 7 (1947), pp. 1034–1041.

———. "History and Organization of a Theater Psychiatric Services Before and After June 30, 1951." In *Recent Advances in Medicine and Surgery (19–30 April 1954): Based on Professional Medical Experiences in Japan and Korea 1950–1953; Vol. II* (Washington, D.C.: Army Medical Service Graduate School, Walter Reed Army Medical Center, Medical Science Publication No. 4, 1954), pp. 358–372.

———. "Lessons Learned." In A. J. Glass and R. J. Bernucci (eds.), *Medical Department, United States Army Neuropsychiatry in World War II, I. Zone of Interior* (Washington, D.C.: U.S. Government Printing Office, 1966), pp. 735–760.

———. "Psychiatry in the Korean Campaign." *United States Army Medical Bulletin,* Vol. 4, No. 10 (1953), pp. 1387–1401.

Hammer, J. "A War's Human Toll: Israel Wins a Fierce Battle, but the Victory Gives Birth to Another Saga of Blood and Fire." *Newsweek,* April 22, 2003.

Hanson, F. "The Factor of Fatigue in the Neuroses of Combat." *United States Army Medical Bulletin,* Vol. 9 (1949), pp. 147–150.

Harma, M., S. Suvanto, S. Popkin, K. Puli, M. Mulder, and K. Hirvonen. "A Dose-Response Study of Total Sleep Time and the Ability to Maintain Wakefulness." *Journal of Sleep Research,* Vol. 7, No. 3 (1998), pp. 167–174.

Helmus, T., L. Rosenthal, C. Bishop, T. Roehrs, M. L. Syron, and T. Roth. "The Alerting Effects of Short and Long Naps in Narcoleptic, Sleep Deprived, and Alert Individuals." *Sleep,* Vol. 20, No. 4 (1997), pp. 251–257.

Iancu, I., O. T. Dolberg, and J. Zohar. "Is Caffeine Involved in the Pathogenesis of Combat-Stress Reaction?" *Military Medicine,* Vol. 161, No. 4 (1996), pp. 230–232.

Ingraham, L. H., and F. J. Manning. "Psychiatric Battle Casualties: The Missing Column in a War Without Replacements." *Military Review*, August 1980, pp. 18–29.

Inzana, C. M., J. E. Driskell, E. Salas, and J. H. Johnston. "Effects of Preparatory Information on Enhancing Performance Under Stress." *Journal of Applied Psychology*, Vol. 81 (1996), pp. 429–435.

The Iowa Persian Gulf Study Group, "Self-Reported Illness and Health Status Among Gulf War Veterans: A Population-Based Study." *Journal of the American Medical Association*, Vol. 277, No. 3 (1997), pp. 238–245.

Johnson, A. W. "Combat Psychiatry, Part 2: The U.S. Army in Vietnam." *Medical Bulletin U.S. Army Europe*, Vol. 25 (1969), pp. 335–39.

Jones, E., and S. Wessely. "Forward Psychiatry in the Military: Its Origins and Effectiveness." *Journal of Traumatic Stress*, Vol. 16, No. 4 (August 2003), pp. 411–419.

Jones, E., and S. Wessely. "Psychiatric Battle Casualties: An Intra- and Interwar Comparison." *British Journal of Psychiatry*, Vol. 178 (2001), pp. 242–247.

Jones, E., K. C. Hyams, and S. Wessely. "Screening for Vulnerability to Psychological Disorders in the Military: An Historical Survey." *Journal of Medical Screening*, Vol. 10 (2003), pp. 40–46.

Jones, E., R. Hodgins-Vermaas, H. McCartney, B. Everitt, C. Beech, et al. "Post-Combat Syndromes from the Boer War to the Gulf War: A Cluster Analysis of Their Nature and Attribution." *BMJ (British Medical Journal)*, Vol. 324 (2002), pp. 321–324.

Kaplan, Z., I. Iancu, and E. Bodner. "A Review of Psychological Debriefing After Extreme Stress." *Psychiatric Services*, Vol. 52, No. 6 (2001), pp. 824–827.

Levav, I., H. Greenfeld, and E. Baruch. "Psychiatric Combat Reactions During the Yom Kippur War." *American Journal of Psychiatry*, Vol. 136, No. 5 (1979), pp. 637–641.

Lewis, B. P., and D. E. Linder. "Thinking About Choking? Attentional Processes and Paradoxical Performance." *Personality and Social Psychology Bulletin*, Vol. 23 (1997), pp. 937–944.

Litz, B. T., M. J. Gray, R. A. Bryan, and A. B. Adler. "Early Intervention for Trauma: Current Status and Future Directions." *Clinical Psychology: Science and Practice,* Vol. 9, No. 2 (2002), pp. 112–134.

Lorch, E. P., D. R. Anderson, and A. D. Well. "Effects of Irrelevant Information on Speeded Classification Tasks: Interference Is Reduced by Habituation." *Journal of Experimental Psychology: Human Perception and Performance,* Vol. 10 (1984), pp. 850–864.

Ludwig, A. O., and S. W. Ranson. "A Statistical Follow-Up of Effectiveness of Treatment of Combat-Induced Psychiatric Casualties: I. Returns to Full Combat." *The Military Surgeon,* January 1947, pp. 51–64.

Madden, D. J. "Aging and Distraction by Highly Familiar Stimuli During Visual Search." *Developmental Psychology,* Vol. 19 (1983), pp. 499–507.

McDuff, D. R., and J. L. Johnson. "Classification and Characteristics of Army Stress Casualties During Operation Desert Storm." *Hospital and Community Psychiatry,* Vol. 43, No. 8 (1992), pp. 812–815.

McNally, R. J., A. B. Richard, and A. Ehlers. "Does Early Psychological Intervention Promote Recovery from Posttraumatic Stress?" *Psychological Science in the Public Interest,* Vol. 4, No. 2 (2003), pp. 45–79.

Mericle, E. W. "The Psychiatric and the Tactical Situations in an Armored Division." *Bulletin of the U.S. Army Medical Department,* Vol. 6, No. 3 (1946), pp. 325–334.

Mullen, B., B. Bryant, and J. E. Driskell. "Presence of Others and Arousal: An Integration." *Group Dynamics: Theory, Research, and Practice,* Vol. 1, No. 1 (March 1997).

Norvikov, V. S. "Psycho-Physiological Support of Combat Activities of Military Personnel," *Military Medical Journal* (Russia), No. 4 (1996), pp. 37–40.

Noy, S. "Battle Intensity and the Length of Stay on the Battlefield as Determinants of the Type of Evacuation." *Military Medicine,* Vol. 152, No. 12 (1987), pp. 601–607.

Noy, S., R. Levy, and Z. Solomon. "Mental Health Care in the Lebanon War, 1982." *Israel Journal of Medical Sciences,* Vol. 20, No. 4 (1984), pp. 360–363.

Plesset, M. R. "Psychoneurotics in Combat." *American Journal of Psychiatry,* Vol. 103 (1946), pp. 87–88.

Rapee, R. M., and R. G. Heimberg. "A Cognitive-Behavioral Model of Anxiety in Social Phobia." *Behaviour Research and Therapy,* Vol. 35, No. 8 (1997), pp. 741–756.

Ritchie, E. C. "Psychiatry in the Korean War: Perils, PIES, and Prisoners of War." *Military Medicine,* Vol. 167, No. 11 (2002), pp. 898–903.

Ritchie, E. C., and D. C. Ruck, "The 528th Combat Stress Control Unit in Somalia in Support of Operation Restore Hope," *Military Medicine,* Vol. 159, No. 5 (1994), pp. 372–376.

Rose, S., J. Bisson, and S. Wessely. "Psychological Debriefing for Preventing Post-Traumatic Stress Disorder (PTSD) (Cochrane Review)." In *The Cochrane Library, Issue 4, 2003* (Chichester, UK: John Wiley & Sons, Ltd.).

Rosenthal, L., T. Roehrs, A. Zwyghuizen-Doorenbos, D. Plath, and T. Roth. "Alerting Effects of Caffeine After Normal and Restricted Sleep." *Neuropsychopharmacology,* Vol. 4, No. 2 (1991), pp. 103–108.

Rosenthal, L., T. A. Roehrs, A. Rosen, and T. Roth. "Level of Sleepiness and Total Sleep Time Following Various Time in Bed Conditions." *Sleep,* Vol. 16, No. 3 (1993), pp. 226–232.

Shephard, B. "Shell-Shock on the Somme." *RUSI Journal,* June 1996, pp. 51–56.

Singer, R. N., J. H. Cauraugh, L. K. Tennant, M. Murphey, et al. "Attention and Distractors: Considerations for Enhancing Sport Performances." *International Journal of Sport Psychology,* Vol. 22 (1991), pp. 95–114.

Solomon, Z., and R. Benbenishty. "The Role of Proximity, Immediacy, and Expectancy in Frontline Treatment of Combat Stress Reaction Among Israelis in the Lebanon War." *American Journal of Psychiatry,* Vol. 143, No. 5 (1986), pp. 613–617.

Solomon, Z., and H. Flum. "Life Events and Combat Stress Reaction in the 1982 War in Lebanon." *Israel Journal of Psychiatry and Related Sciences,* Vol. 23, No. 1 (1986), pp. 9–16.

Solomon, Z., and H. Flum. "Life Events, Combat Stress Reaction and Post-Traumatic Stress Disorder." *Social Science and Medicine,* Vol. 26, No. 3 (1988), pp. 319–325.

Solomon, Z., R. Benbenishty, and M. Mikulincer. "A Follow-Up of Israeli Casualties of Combat Stress Reaction ('Battle Shock') in the 1982 Lebanon War." *British Journal of Clinical Psychology,* Vol. 27, Pt. 2 (1988), pp. 125–135.

Solomon, Z., M. Mikulincer, and S. E. Hobfoll. "Effects of Social Support and Battle Intensity on Loneliness and Breakdown During Combat." *Journal of Personality and Social Psychology,* Vol. 51, No. 6 (1986), pp. 1269–1276.

Solomon, Z., S. Noy, and R. Bar-On. "Risk Factors in Combat Stress Reaction: A Study of Israeli Soldiers in the 1982 Lebanon War." *Israel Journal of Psychiatry and Related Sciences,* Vol. 23, No. 1 (1986), pp. 3–8.

Steiner, M., and M. Neumann. "Traumatic Neurosis and Social Support in the Yom Kippur War Returnees." *Military Medicine,* Vol. 143, No. 12 (1978), pp. 866–868.

Thyer, B. A., et al. "In Vivo Distraction—Coping in the Treatment of Test Anxiety." *Journal of Clinical Psychology,* Vol. 37 (1981), pp. 754–764.

van Emmerik, A. A., J. H. Kamphuis, A. M. Hulsbosch, and P. M. Emmelkamp. "Single Session Debriefing After Psychological Trauma: A Meta-Analysis." *Lancet,* Vol. 360, No. 9335 (2002), pp. 766–771.

Vossel, G., and L. Laux. "The Impact of Stress Experience on Heart Rate and Task Performance in the Presence of a Novel Stressor." *Biological Psychology,* Vol. 6, No. 3 (1978), pp. 193–201.

Wessely, S. "Psychological Debriefing Is a Waste of Time," *British Journal of Psychiatry,* Vol. 183 (2003), pp. 12–14.

Wine, J. "Test Anxiety and Direction of Attention." *Psychological Bulletin,* Vol. 76 (1971), pp. 92–104.

Yitzhaki, T., Z. Solomon, and M. Kotler. "The Clinical Picture of Acute Combat Stress Reaction Among Israeli Soldiers in the 1982 Lebanon War." *Military Medicine,* Vol. 156, No. 4 (1991), pp. 193–197.

Books

Ahrenfeldt, R. H. *Psychiatry in the British Army in the Second World War.* London: Routledge & Kegan Paul Ltd., 1958.

American Psychiatric Association. *Diagnostic and Statistical Manual for Mental Disorders*. 4th ed. Washington, D.C.: American Psychiatric Association, 1994.

Appel, J. W. "Preventive Psychiatry." In R. S. Anderson, A. J. Glass, and R. J. Bernucci (eds.), *Medical Department, United States, Neuropsychiatry in World War II. Vol. I: Zone of the Interior*. Washington, D.C.: U.S. Government Printing Office, 1966, pp. 373–414.

Ashworth, G. J. *War and the City*. New York: Routledge, 1991.

Babington, A. *Shell-Shock: A History of the Changing Attitudes to War Neurosis*. London: Leo Cooper, 1997.

Bean, C.E.W. *Official History of Australia in the War of 1914–1918. Vol. 3: The AIF in France 1916*. Canberra: Australian War Memorial, 1929.

Belenky, G. L., S. Noy, and S. D. Solomon. "Battle Stress, Morale, Cohesion, Combat Effectiveness, Heroism, and Psychiatric Casualties: The Israeli Experience." In G. L. Belenky (ed.), *Contemporary Studies in Combat Psychiatry*. Westport, CT: Greenwood Press, Inc., 1987.

Blumenson, M. *The European Theater of Operations, Breakout and Pursuit*. Washington, D.C.: Center of Military History, United States Army, 1989.

Bradley, J., and R. Powers. *Flags of Our Fathers*. New York: Bantam Books, 2000.

Brill, N. Q., and G. W. Beebe. *A Follow-Up Study of War Neuroses*. Washington, D.C.: VA Medical Monograph, 1955.

Charlton, P. *Pozieres*. London: Leo Cooper, 1986.

Clausewitz, C. *On War*. Michael Howard and Peter Paret (eds. and trans.). Princeton, NJ: Princeton University Press, 1984.

Drayer, C. S., and A. J. Glass. "Introduction." In A. J. Glass (ed.), *Medical Department, United States Army, Neuropsychiatry in World War II. Vol. 2: Overseas Theaters*. Washington D.C.: Office of the Surgeon General, Department of the Army, 1973.

Driskell, J. E., and J. H. Johnston. "Stress Exposure Training." In J. A. Cannon-Bowers and E. Salas (eds.), *Making Decisions Under Stress: Implications for Individual and Team Training*. Washington, D.C.: American Psychological Association, 1998, pp. 191–217.

Dubberly, B. C. "Drugs and Drug Use." In S. C. Tucker (ed.), *Encyclopedia of the Vietnam War: A Political, Social and Military History,* Santa Barbara: ABC-CLIO, 1998, pp. 179–180.

Ellis, J. *Eye-Deep in Hell.* London: Croom Helm, 1976.

Glenn, R. W. *Reading Athena's Dance Card.* Annapolis, MD: Naval Institute Press, 2000.

Glieg, G. R. *The Subaltern: A Chronicle of the Peninsular War.* Edinburgh, 1877.

Grinker, R. R., and J. P. Spiegel. *Men Under Stress.* New York: McGraw Paperbacks, 1963.

Haslam, D. R., and P. Abraham. "Sleep Loss and Military Performance." In G. L. Belenky (ed.), *Contemporary Studies in Combat Psychiatry.* Westport, CT: Greenwood Press, 1987, pp. 167–184.

Hessin, A. L. "Neuropsychiatry in Airborne Divisions." In W. S. Mullins and A. J. Glass (eds.), *Medical Department, United States Army Neuropsychiatry in World War II. Vol. 2: Overseas Theaters.* Washington, D.C.: U.S. Government Printing Office, 1973, pp. 375–398.

Holmes, R. *Redcoat: The British Soldier in the Age of Horse and Musket.* London: HarperCollins Publishers, 2001.

Janis, I. L. *Air War and Emotional Stress.* New York: McGraw-Hill, 1951.

Jones, F. D. "Psychiatric Lessons of War." In F. D. Jones, L. R. Sparacino, V. L. Wilcox, J. M. Rothberg, and J. W. Stokes (eds.), *War Psychiatry.* Washington, D.C.: TMM Publications, 1995a, pp. 1–33.

Jones, F. D. "Traditional Warfare Combat Stress Casualties." In F.D. Jones et al. (eds.), *War Psychiatry*, Washington, D.C.: TMM Publications, 1995b, pp. 35–61.

Kellett, A. *Combat Motivation.* Boston, The Hague, London: Kluwer Nijhoff Publishing, 1982.

Kulka, R. A., W. E. Schlenger, J. A. Fairbank, R. L. Hough, B. K. Jordan, et al. *Trauma and the Vietnam War Generation.* New York: Brunner/Mazel, 1990.

McCallum, J. "Medicine, Military." In S. C. Tucker (ed.), *Encyclopedia of the Vietnam War: A Political, Social and Military History,* Santa Barbara: ABC-CLIO, 1998, pp. 423–428.

Noy, S. "Stress and Personality as Factors in the Causation and Prognosis of Combat Reactions." In G. L. Belenky (ed.), *Contemporary Studies in Combat Psychiatry.* Westport, CT: Greenwood Press, 1987, pp. 21–30.

Shephard, B. *A War of Nerves.* London: Jonathan Cape, 2000.

Shipp, J. *The Path of Glory.* London, 1969.

Sledge, E. B. *With the Old Breed at Peleliu and Okinawa.* New York: Oxford University Press, 1981.

Smith, R. R. *The War in the Pacific: Triumph in the Philippines.* Washington, D.C.: Center of Military History, United States Army, 1991.

Solomon, Z., N. Laor, and A. C. McFarlane. "Acute Posttraumatic Reactions in Soldiers and Civilians." In B. A. van der Kolk, A. C. McFarlane, and L. Weisaeth (eds.), *Traumatic Stress: The Effects of Overwhelming Experience on Mind, Body, and Society.* New York, London: The Guilford Press, 1996.

Stouffer, S. A., A. A. Lumsdaine, M. H. Lumsdaine, et al. *The American Soldier, Combat and Its Aftermath.* New York: John Wiley & Sons, Inc., 1965.

Strang, William. *Diaries.* Liddle Collection, University of Leeds.

Thompson, L. J., P. C. Talkington, and A. O. Ludwig. "Neuropsychiatry at Army and Division Levels." In W. S. Mullins and A. J. Glass (eds.), *Medical Department, United States Army Neuropsychiatry in World War II. Vol. 2: Overseas Theater.* Washington, D.C.: U.S. Government Printing Office, 1973. pp. 275–374.

Tzu, S. *The Art of War.* Samuel B. Griffith (trans.). New York: Oxford University Press, 1982.

Ursano, R. J., T. A. Grieger, and J. E. McCarroll. "Prevention of Posttraumatic Stress: Consultation, Training, and Early Treatment." In B. A. van der Kolk, A. C. McFarlane, and L. Weisaeth (eds.),*Traumatic Stress: The Effects of Overwhelming Experience on Mind, Body and Society.* New York: The Guilford Press, 1996.

Warr, N. *Phase Line Green: The Battle for Hue, 1968.* New York: Ivy Books, 1997.

Watson, P. *War on the Mind: The Military Uses and Abuses of Psychology.* New York: Basic Books, 1978.

White, W. L. *Back Down the Ridge*. New York: Harcourt, Brace and Company, 1953.

Military and Technical Reports

After Action Report, Medical Section, Ninth U.S. Army, Period 5–30 September 1944 Inclusive. National Archives and Records Administration. Record Group 407.

Alexander, J. H., ed. *Battle of the Barricades, U.S. Marines in the Recapture of Seoul.* Washington, D.C.: Marine Corps Historical Center, 2000.

Annex Queen to 1st Marine Division Special Action Report, Division Surgeon, August 28 to October 7. Marine Corps History and Museum Division, Korean War, CD #1.

Avidor, Gideon [BG, IDF, ret.]. Presentation to members of the 10th Mountain Division command and staff, January 13, 2003, Fort Drum, New York.

Battle Fatigue: Normal Common Signs, What to Do for Self and Buddy. Washington, D.C.: Headquarters, Department of the Army, GTA 21-3-4, June 1986, http://www.bragg.army.mil/528CSC/GTA21-3-4.htm (last accessed November 21, 2003).

Battle Fatigue: Warning Signs; Leader Actions. Washington, D.C.: Headquarters, Department of the Army, GTA 21-3-5, June 1994, http://www.bragg.army.mil/528CSC/GTA21-3-5.htm (last accessed November 21, 2003).

Beebe, G. W., and J. W. Appel. *Variation in Psychological Tolerance to Ground Combat in World War II.* Washington, D.C.: National Academy of Sciences, 1958.

Belenky, G. L. *Sleep, Sleep Deprivation, and Human Performance in Continuous Operations.* http://www.usafa.af.mil/jscope/JSCOPE97/Belenky97/Belenky97.htm (last accessed November 21, 2003).

Belenky, G. L., C. F. Tyner, and F. J. Sodetz. *Israeli Battle Shock Casualties: 1973 and 1982.* Washington, D.C.: Walter Reed Army Institute of Research, 1983.

Besser, Y. *Military Operations in Urbanized Terrain—Medical Aspects—Lebanon War 1982, A Case Study*. Unpublished paper, August 1985.

Corregidor Then and Now, Battle of Manila, http://corregidor.org/chs_manila/mjump.htm (accessed Sept 12, 2003).

Department of Defense (DoD). *Combat Stress.* Department of the Army, Field Manual (FM) 6-22.5, June 23, 2000.

———. "Chapter 4: Combat Stress Control Consultation." In *Combat Stress Control in a Theater of Operations*. Department of the Army, Field Manual (FM) 8-51, http://www.vnh.org/FM851/chapter4.html (last accessed May 22, 2004).

———. *Leaders' Manual for Combat Stress Control.* Department of the Army, Field Manual (FM) 22-51, http://www.vnh.org/FM22-51/booklet1.html (last accessed May 22, 2004).

Eversmann, M. "The Urban Area During Support Missions Case Study: Mogadishu; The Tactical Level I." In R. W. Glenn (ed.), *Capital Preservation: Preparing for Urban Operations in the Twenty-First Century; Proceedings of the RAND Arroyo-TRADOC-MCWL-OSD Urban Operations Conference, March 22–23, 2000.* Santa Monica, CA: RAND Corporation, CF-162-A, 2001.

Glass, A. J. "Leadership Problems of Future Battle: Presented to The U.S. Army War College." Carlisle Barracks, Pennsylvania, 1959.

Glenn, R. W. *AAR from the Warlords [24 Marine Expeditionary Unit].* May 7, 2003.

———. *Combat in Hell: A Consideration of Constrained Urban Warfare.* Santa Monica. CA: RAND Corporation, MR-780-A/DARPA, 1996.

Glenn, R. W., S. L. Hartman, and S. Gerwehr. *Urban Combat Service Support Operations: The Shoulders of Atlas*. Santa Monica, CA: RAND Corporation, MR-1717-A, 2003.

Glenn, R. W., S. W. Atkinson, M. P. Barbero, F. J. Gellert, S. Gerwehr, S. L. Hartman, J. J. Medby, A. W. O'Donnell, D. Owen, and S. Pieklik. *Ready for Armageddon: Proceedings of the 2001 RAND Arroyo-Joint ACTD-CETO-USMC Nonlethal and Urban Operations Program Urban Operations Conference.* Santa Monica, CA: RAND Corporation, CF-179-A, 2002.

Hall, M. T., and M. T. Kennedy. "The Urban Area During Support Missions Case Study: Mogadishu; Applying the Lessons Learned: Take 2." In R. W. Glenn (ed.), *Capital Preservation: Preparing for Urban Opera-*

tions in the Twenty-First Century; Proceedings of the RAND Arroyo-TRADOC-MCWL-OSD Urban Operations Conference March 22–23, 2000. Santa Monica, CA: RAND Corporation, CF-162-A, 2001.

Headquarters, First Medical Battalion, APO 1, United States Army. National Archives and Records Administration. Record Group 407.

Headquarters, 29th Infantry Division. National Archives and Records Administration, Record Group 407.

History of the 18th Infantry (1st ID) for the Period 1–31 October 1944. National Archives and Records Administration. Record Group 407.

Huber, T. M. *The Battle of Manila*. Fort Leavenworth, KS: Combat Studies Institute, Command and General Staff College, http://www.globalsecurity.org/military/library/report/2002/MOUTHuber.htm (accessed September 22, 2003).

Lawrence, C., and R. Anderson. *Measuring the Effects of Combat in Cities: Phase 1*. Annandale, VA: The Dupuy Institute, 2002.

Marlowe, D. H. *Cohesion, Anticipated Breakdown, and Endurance in Battle: Considerations for Severe and High Intensity Combat*. Washington, D.C.: Walter Reed Army Institute of Research, 1979.

———. *Psychological and Psychosocial Consequences of Combat and Deployment with Special Emphasis on the Gulf War*. Santa Monica, CA: RAND Corporation, MR-1018/11-OSD, 2001.

Marshall, S.L.A. "Combat Leadership." In *Preventive and Social Psychiatry*,. Washington, D.C.: U.S. Government Printing Office, 1957, pp. 303–307.

Neel, S. *Vietnam Studies: Medical Support of the U.S. Army in Vietnam, 1965–1970*. Washington, D.C.: U.S. Department of the Army, 1973.

Report of the Division Psychiatrist, 37th ID, Luzon Campaign. National Archives and Records Administration, Record Group 407.

Study of AGF Battle Casualties. Washington, D.C.: Headquarters, Army Ground Forces Plans Section, 1946.

Thayer, T. C. (ed.) *A Systems Analysis View of the Vietnam War*. Alexandria, VA: Defense Technical Information Center, 1978.

Thomas, T. L. and C. P. O'Hara, "Combat Stress in Chechnya: The Equal Opportunity Disorder," Fort Leavenworth, KS: Foreign Military Studies

Office, http://fmso.leavenworth.army.mil/fmsopubs/issues/stress.htm (accessed December 2, 2004).

U.S. Army Surgeon General and HQDA G1, *Operation Iraqi Freedom Mental Health Advisory Team Report,* December 16, 2003, http://www.globalsecurity.org/military/library/report/2004/mhat_report.pdf (accessed May 15, 2004).

Willbanks, J. H. *The Battle for Hue, 1968*. Fort Leavenworth, KS: Combat Studies Institute, Command and General Staff College, 2002, http://www.globalsecurity.org/military/library/report/2002/MOUTWilbanks.htm (accessed September 20, 2003).

Wright, K. M., D. H. Marlowe, J. A. Martin, R. K. Gifford, G. L. Belenky, and F. J. Manning. *Operation Desert Shield Desert Storm: A Summary Report*. Washington, D.C.: Walter Reed Army Institute of Research, 1995.

Yates, L. A. *Operation Just Cause in Panama City, December 1989*. Fort Leavenworth, KS: Combat Studies Institute, Command and General Staff College, 2002.

Film

Black Hawk Down, directed by Ridley Scott, Columbia Pictures, 2001.

Interviews

Allison, John [LtCol, USMC, ret.]. Interview with the author, Arlington, Virginia, June 5, 2003. LtCol Allison served in Operation Restore Hope, Mogadishu, Somalia, following service in the first Persian Gulf War where he was wounded. He also participated in peacekeeping operations in Los Angeles, California.

Barrow, Robert [LtCol, USMC, ret.]. Interview with the author, Tampa, Florida, June 16, 2003. LtCol Barrow served in Beirut, Lebanon, the Gulf War, and Mogadishu, Somalia.

Belenky, Gregory, [M.D., COL, USA MC]. Interview with the author, Washington, D.C., May 20, 2003. Dr. Belenky is a psychiatrist. He is the director of Neuropsychiatry at the Walter Reed Army Institute of Research and was a mental health officer in the 2nd Armored Cavalry Regiment.

Brock, J. Price [M.D., LCDR, USN MC, ret.]. Interview with the author, Abilene, Texas, March 6, 2003. LCDR Brock was the surgeon for the 1/5 Marines during the battle of Hue. He is currently a semi-retired orthopedic surgeon.

Brooks, Johnny W. [COL, USA, ret.]. Written comment to the author, May 6, 2003. COL Brooks was the commander of the 4th Battalion, 17th Infantry during Operation Just Cause.

Castro, Carl [Ph.D., LTC, USA MC]. Interview with the author, Washington, D.C., May 20, 2003. LTC Castro is a psychologist who has served in Bosnia and Kosovo. Presently he is a research scientist at the Walter Reed Army Institute of Research.

Christmas, George [LtGen, USMC, ret.]. Written comment to the author, June 6, 2003. LtGen Christmas was the company commander of Fox/2/5 during the battle of Hue. His actions in this battle earned him the Navy Cross.

———. Interview with the author, Stafford, Virginia, February 28, 2003.

Collyer, Robert [COL, Australian Army]. Interview with the author, Brisbane, Australia, March 7, 2003. COL Collyer is a psychologist with the Australian Army. He has operational service in East Timor and Bougainville and he presently does human factors work related to urban operations, Army structure, and negligent weapon discharges.

Dolev, Eran [M.D., BG, IDF, ret.]. Interview with the author, London, England, March 20, 2003. BG Dolev was the Surgeon General of the Israeli Defense Forces during the 1982 Lebanon War.

Downs, Michael [BGen, USMC, ret.]. Interview with the author, Quantico, Virginia, February 12, 2003. BGen Michael Downs was the company commander of Fox/2/5 during the battle of Hue.

Doyle, Michael E. [M.D., MAJ, USA MC]. Interview with the author, Tacoma, Washington, February 20, 2003. MAJ Doyle presently serves as Chief of Outpatient Psychiatry, Madigan Army Medical Center, Tacoma, Washington. He also commanded the 98th Medical Detachment of Task Force Eagle during deployment to Bosnia-Herzegovina.

Eversmann, Matthew [MSG, USA]. Interview with the author, Baltimore, Maryland, May 22, 2003. MSG Eversmann served with Bravo Company, 3/75 Ranger Regiment in Task Force Ranger, Mogadishu, Somalia.

Greenberg, Neil [M.D., LtCdr, Royal Navy]. Interview with the author, London, England, May 5, 2003. LtCdr Greenberg is a psychiatrist with the Royal Navy. He is presently a Specialist Registrat in Liaison Psychiatry at Maudsley Hospital, London.

Harrington, Myron [Col, USMC, ret.]. Interview with the author, Mount Pleasant, South Carolina, February 24, 2003. Col Harrington was the company commander of Delta/1/5 during the battle of Hue.

Holcomb, John [M.D., COL, USA]. Interview with the author, Houston, Texas, March 4, 2003. Dr. Holcomb is a trauma surgeon who served in Operation Just Cause and in Somalia with Task Force Ranger. He is presently a consultant for the Surgeon General and commands the Army's Institute for Surgical Research.

Holloway, Harry [M.D., COL, USA MC, ret.]. Written comments and interview with the author, Bethesda, Maryland, May 22, 2003. Dr. Holloway presently serves as Professor of both Psychiatry and Neuroscience at the Uniformed Services University of Health Sciences.

Keller, Richard [MAJ, USA MC]. Interview with the author, Washington, D.C., May 21, 2003. MAJ Keller was a member of the Deployment Cycle Support Program and subsequently participated with the Mental Health Advisory Team (MHAT).

Lawrence, Kevin [WO2, USMC]. Interview with the author, Washington, D.C., July 1, 2003. WO2 Lawrence has served in Operation Stabilize in East Timor and Operation Restore Hope in Mombasa, Kenya.

Levy, Ron [Psy.D., COL, IDF, ret.]. Interview with the author, Ksarsaba, Israel, April 7, 2003. COL Levy was the chief psychologist for the Israeli Army during the 1982 Lebanon War.

Lewis, Larry K. [Ph.D., COL, USA MC]. Interview with the author, Fort Meade, Maryland, August 21, 2003. COL Lewis was the command psychologist with Task Force Ranger during operations in Somalia. He currently works with the 902nd Military Intelligence Group.

Litz, Brett T. [Ph.D.]. Interview with the author, Boston, Massachusetts, May 19, 2003. Dr. Litz is the Associate Director for the National Center for Post-Traumatic Stress Disorder and Professor at Boston University.

———. Written comments to the author, December 18, 2004.

Marlowe, David H. [Ph.D.]. Interview with the author, Alexandria, Virginia, May 21, 2003. Dr. Marlowe is a medical anthropologist who worked for the Walter Reed Army Institute of Research, where he was Chief of the Department of Military Psychiatry for over 20 years. He is currently a senior lecturer in Psychiatry at the Uniformed Services University of Health Sciences.

Meadows, Chuck [Col, USMC, ret.]. Interview with the author, Bainbridge, Washington, February 18, 2003. Col Meadows was company commander of Golf/2/5 during the battle of Hue.

Morgan, Charles A. [M.D.]. Interview with the author, New Haven, Connecticut, March 27, 2003. Dr. Morgan is an Associate Professor of Psychiatry at Yale University.

Mosebar, Robert H. [M.D.]. Interview with the author, San Antonio, Texas, March 13, 2003. Dr. Mosebar served in the Army for 34 years with operational experience in World War II, the Korean War, and the Vietnam War. During World War II he participated in the battle of Manila.

Pierce, Jack S. [M.D., CDR, USN MC]. Interview with the author, Arlington, Virginia, May 20, 2003. CDR Pierce was the psychiatrist for the Marine Corps 2nd Division and is currently the Clinical Programs Officer for the Marine Corps.

Ritchie, Elspeth C. [M.D., COL, USA MC]. Interview with the author, Washington, D.C., June 3, 2003. LTC Ritchie served with the 528 Combat Stress Control Unit during Operation Restore Hope. She presently works at the Uniformed Services University of the Health Sciences.

———. Written comments to the author, December 14, 2004.

Shalev, Arieh [M.D.]. Interview with the author, Jerusalem, Israel, July 7, 2003. Dr. Shalev was a psychiatrist with the Israeli Defense Force Medical Corps. He conducts research in the field of stress reactions.

Stokes, James [M.D., COL, USA MC]. Interview with the author, Fort Sam Houston, Texas, March 19, 2003. COL Stokes serves as the Combat and Operational Stress Control Program Officer for the U.S. Army's Medical Corps.

———. Written comments to the author, November 18, 2004.

Thompson, Bob [Col, USMC, ret.]. Interview with the author, Fredericksburg, Virginia, February 25, 2003. Col Thompson was the battalion commander of 1/5 during the battle of Hue.

Warr, Nicholas. Interview with the author, Solana Beach, California, May 9, 2003. Nicholas Warr was the platoon commander of C/1/5 during the battle of Hue. He is also the author of *Phase Line Green: The Battle for Hue, 1968.*

Wessely, Simon [M.D.]. Interview with the author, London, England, May 2, 2003. Dr. Wessely is Professor of Epidemiological and Liaison Psychiatry at Kings College London.